The Fruit Hunters

A Story of Nature, Adventure, Commerce and Obsession

Adam Leith Gollner

SCRIBNER

New York   London   Toronto   Sydney

SCRIBNER
A Division of Simon & Schuster, Inc.
1230 Avenue of the Americas
New York, NY 10020

First Scribner hardcover edition May 2008

SCRIBNER and design are trademarks of
The Gale Group, Inc., used under license
by Simon & Schuster, the publisher of this work.

For information about special discounts for bulk purchases,
please contact Simon & Schuster Special Sales:
1-800-456-6798 or business@simonandschuster.com

Book design by Ellen R. Sasahara
Text set in Aldine 401 BT

Manufactured in the United States of America

1 3 5 7 9 10 8 6 4 2

Library of Congress Cataloging-in-Publication Data
Gollner, Adam Leith, 1976–
The fruit hunters : a story of nature, adventure, commerce and obsession /
Adam Leith Gollner.
p.   cm.
1. Fruit. 2. Tropical fruit. 3. Fruit culture. 4. Fruit trade. I. Title.
SB354.8.G65 2008
641.3'4—dc22   2007042423

ISBN-13: 978-0-7432-9694-6
ISBN-10:   0-7432-9694-X

For Liane

Here is always spring, and every fruit;
this is the nectar of which each tells.

—Dante, *Purgatorio*

Contents

Part 4 Obsession • 213

Prologue

Blame It on Brazil

It is here that we harvest the miraculous fruits your heart hungers for;
come and intoxicate yourself on the strange sweetness.

—Charles Baudelaire, *The Voyage*

WIPING SAND from my eyes, I stumble off a bus outside the Rio de Janeiro botanical garden and pass under the Ionic columns at the entrance. A dirt road leads to the greenery. Royal palms line the way, cathedral pillars vaulting into a canopy.

A fuzzy, neon-green hot dog slithers across my path. I start taking photographs of the creature—a giant millipede—as it undulates toward a plastic orange trash can warping in the heat. Getting deeper into the garden, I come upon the bust of some forgotten botanist, a droplet of tree sap trickling down his forehead like a misplaced tear.

I rest on a bench near a lagoon strewn with lily pads. The silhouette of Cristo Redentor looms down from the summit of Mount Corcovado. Rio isn't quite the fantasyland of Bossa Nova melodies and paradisiacal seascapes I had envisioned. Homeless kids sleep facedown on Ipanema's wavy mosaic boardwalks. Blue smoke curls over rivulets of shantytown sewage. The only good photograph I've taken is of a black dog lying on the beach at dusk, an ominous canine stain surrounded by white sand, turquoise water and a purple-pink twilight.

I try to not think about home. My grandfather just died. My parents' marriage is dissolving. A loved one's manic depression is spiraling into a gruesome battle with addiction. My best friend, recovering from a suicide attempt, has been diagnosed with paranoid schizophrenia. To top it all off, my girlfriend of eight years is spending New Year's in Europe with her new lover, a French soldier.

I hear a mocking cackle in the foliage above, and spy a couple of toucans kissing each other's Technicolor beaks. Suddenly, a nearby tree shakes with commotion. Two ring-tailed, white-whiskered monkeys are playing tag. After a momentary stare-down, one of the monkeys plunges through the air into another tree. Zooming in through my viewfinder, I notice something odd: the branches appear to be sprouting bran muffins.

I pick one of the muffins off the ground. It's brown and woody. It feels like it was baked in a buttered tray at 350 degrees for two hours too long. Not only is the muffin rock hard, it's also hollowed out, as though someone had flipped it over and scooped out its insides. The shell's interior bears scratch marks and a couple of fibrous veins. I wonder what was once inside these empty confections.

A plaque identifies the tree as a sapucaia. In season, the cupcakes grow packed with a half dozen seeds shaped like orange segments. At ripeness, these burst through the base, scattering on the ground. Impatient young monkeys sometimes punch into an unripe muffin and wrap their fingers around a fistful of nuts. Because their cognitive faculties are not developed enough to understand that extracting their paws requires letting go of the nuts, they end up dragging their sapucaia handcuffs around for miles.

In English, these sapucaias are called paradise nuts, an appellation dating back to the European discovery of the New World, then considered the site of heaven. In the sixteenth century, France's Jean de Lery became convinced that he had found Eden in a Brazilian pineapple patch. In 1560, Portuguese explorer Rui Pereira announced that Brazil was officially paradise on earth.

If I can't find paradise in Brazil, maybe I can find some paradise nuts instead. I head to a small grocery market outside the park. Any sapucaias? The cashier shakes his head, but offers me a Brazil nut, which he says is similar. Biting into it, I'm amazed at how creamy and coconutty it is compared to those impossible-to-open monstrosities that lurked in childhood Christmas nut bowls.

Oversized pineapples, melons and clusters of bananas hang from the ceiling in mesh netting. I pick up a cashew apple, which looks a lot like an angry red pepper capped with a crescent-shaped nut. The green cambucis resemble miniature B-movie flying saucers. The billiard-ball-sized guavas are so fragrant that the ones I buy perfume my hotel room for the rest of my stay.

HEADING TOWARD the beach, I eat my way through a shopping bag full of the salesman's untranslatable recommendations. Sinking my teeth into one of the scarlet pearlike *jambos* makes me think of crunching on refreshingly sweet Styrofoam. The transparent gummy flesh of a lemon-shaped *abiu* tastes like a cross between wine gums and the crème caramels served in French bistros back home. A machete-toting coconut vendor, noticing my tentative nibbling of a *maracujá's* bitter skin, slices the orb in half and shows me how to slurp up the lavender-fruit-punch viscera.

I enter a *suco* bar, one of the countless juice stalls brightening Rio's crumbling street corners, wondering if I can recognize any of the fruits. The purple *açai* berries on the menu look like the marbles we called "nightmares" in grade two. Across the counter sits a crate full of eyeballs. The owner hands me one of the red-rimmed ocular globules, and out dangles an optic nerve attached to a pitch-black iris and a leering white sclera. It's a *guaraná* fruit, he says, a natural stimulant that's processed to make energy-boosting shakes and soft drinks. I stare at it staring back at me.

Hypnotized, I copy the list of fruits on the stand's menu into my notebook.

By now, the sun is setting into the pastel horizon. Clouds of confetti swirl through the air, paving the ground with tropical snow. I nearly forgot that it's New Year's Eve. The beach has filled with revelers dressed in white. Many have come to the ocean from hillside slums, bearing statues of Macumba saints. They light candles and arrange bits of ribbon around sacrificial flower petals for Iemanjá, the shape-shifting spirit of the waters. As their prayers crash against the waves, the surface of the sea dances with offerings.

I look down at my list of fruits and recite the names under my breath, syncing into the rhythms of nearby batucada drummers. Softly chanting, I close my eyes, and feel a sense of peace. For a moment, I forget everything. I forget my name. I forget why I came here. All I know is *abacaxí, açai, ameixa, cupuaçu, graviola, maracujá, taperebá, uva, umbu.*

Introduction

The Fruit Underworld

Man, you know Adam enjoyed things that kings and
queens will never have and things kings and queens can't
never get and they don't even know about.

—Howlin' Wolf, *Going Down Slow*

T HERE IS A theory that explains humankind's communion with
fruits: biophilia, or the "love of life." Psychologist Erich Fromm
coined the term in 1964 as a way of describing the innate attraction
to processes of life and growth. The hypothesis suggests that organisms facing death can preserve themselves through contact with living systems.
Biologists then adopted the term, noting a tendency for humans to feel a
spiritually transformative connectedness with nature. "Our existence
depends on this propensity," wrote Harvard entomologist Edward O.
Wilson. Citing evidence of quicker rates of recovery for patients exposed to
images of green spaces, scientists speculate that biophilia is an evolutionary
mechanism ensuring the survival of interdependent life-forms.

In Brazil, fruits seemed to be calling out to me. I returned the call.
From then on, I couldn't seem to shake them.

As mundane as they may appear, fruits are also deeply alluring. To
begin with, there's something unusual about their very omnipresence.
Fruits are everywhere, perspiring on street corners, chilling in hotel lob-
bies and on teachers' desks, coagulating in yogurts and drinks, adorning
laptops and museum walls.

Although a select few species dominate international trade, our whole planet is brimming with fruits that are inaccessible, ignored and even forbidden. There are mangoes that taste like piña coladas. Orange cloudberries. White blueberries. Blue apricots. Red lemons. Golden raspberries. Pink cherimoyas. Willy Wonka's got nothing on Mother Nature.

The diversity is dizzying: most of us have never heard of the *araça,* but Amazonian fruit authorities say there are almost as many types of this yellow-green guava relative as there are beaches in Brazil. Within the tens of thousands of edible plant species, there are hundreds of thousands of varieties—and new ones are continually evolving. Magic beans, sundrops, cannonballs, delicious monsters, zombi apples, gingerbread plums, swan egg pears, Oaxacan trees of little skulls, Congo goobers, slow-match fruits, candle fruits, bastard cherries, bignays, belimbings, bilimbis and biribas. As Hamlet might've said: "There are more fruits in heaven and earth, Horatio, than are dreamt of in your philosophy."

Among the fruit world's most euphonious offerings, the clove lilly pilly tastes like pumpkin pie and goes well with kangaroo meat. Existentialists might prefer camu-camus, which are purple drops of sour deliciousness. The yum-yum tree sprouts what appear to be fluffy yellow dusters. Certain Pacific islands have yang-yang trees up the yin yang. Other fruity two-twos include far-fars, lab-labs, num-nums, jum-lums and lovi-lovis.

Many botanically documented plants, like the looking-glass tree, appear to have somehow escaped from a Lewis Carroll laudanum reverie. The pincushion fruit, with its spiked cloak of white rays, is like an exploding star frozen in time. The toothbrush tree's fruits are eaten before bed in the Punjab and the fruits of the toothache tree are used in Virginia to alleviate dental malaise. Succulent umbrella fruits are cherished in the Congo. The glistening puddinglike eta fruit is eaten by tilting the head back and slurping it down like an oyster. The fruits of the toad tree look like frogs and taste like carrots. The milk orange of Wen-chou is a citrus fruit shot though with a creamy mist that, when peeled, swirls enchantingly through the air. Kids play football with the fruit of the money tree. The emu apple is eaten after being buried in soil for several days. Sword fruits call to mind dangling sabers in the moonlight; they're also called broken bones plants or midnight horrors because clumps of fallen fruit are periodically mistaken for skeletal remains.

The pirate books young children devour occasionally mention the inconceivably delicious fruits that buccaneers used to eat while hiding out on tropical islands. In Neverland, the Lost Boys and Peter Pan, "clad in skeleton leaves and the juices that ooze out of trees," ate roasted bread-fruits, mammee apples and calabashes of poe-poe. It was only in Brazil that I realized such fruits were real. There are thousands upon thousands of fruits that we never imagined—and that few of us will ever taste, unless we embark on fruit-hunting expeditions.

In the tropics, kids eat rare jungle fruits the way North Americans eat candy. Even fruits that we've learned to steer clear of at supermarkets suddenly taste excellent in their native lands. When I first encountered a papaya on a teenaged trip to Central America, I was astounded by its flavor, how it filled my mouth with an edible perfume. The ones at home all tasted vaguely unhygienic.

In my experience, fruits are inextricably linked with travel, with other lands, with escaping. Growing up in suburban Montreal, winters were pretty fruitless. When I was thirteen, my family moved to Budapest for a couple of years. My brothers and I had never tasted apricots, peaches and tomatoes as good as the ones that grew in our backyard and in our relatives' orchards. It was easy to see why the Hungarian word for "paradise" also means tomato: *paradiscom.*

Ten years later, I tasted a grape at my father's Hungarian vineyard that floored me with a recollection from age four or five. It was dawn, and my brother and I woke up to go buy grape Bubblicious at the Black Cat, a candy store down the street. The store was off-limits—as was most candy—but, overcome with desire for those cubes of purple awesomeness, we decided the solution was simply to go and get some before our parents woke up. We arrived at the Black Cat as the sun was rising. Needless to say, it wasn't open. We peered through the window at the fireworks, comic books, arcade games, and all those candies. Clutching our fistfuls of nickels and dimes, we hiked back home to our anxious parents, who had called the police and started a manhunt. Like something out of Ingmar Bergman's *Wild Strawberries,* the buried escapade returned with an instant recall the moment I tasted that Concord grape.

Pablo Neruda said that when we bite into apples, we become, for an instant, young again. When I was in Paris, an Algerian taxi driver spent the entire ride describing the prickly pears of his youth, lamenting their taste in France but vividly recalling how sweet they were in his home-land. A wholesaler in New York told me of discovering a quince perfuming the clothes in his mother's armoire when he was a child. "What did you do when you found it?" I asked. "I sniffed it," he replied.

Bertolt Brecht once wrote a poem about seeing some fruits in the tree outside his window that teleported him to a more innocent age. In the verses he spends a few minutes debating quite seriously whether to put on his glasses "in order to see those black berries again on their tiny red stalks." The poem ends without any resolution. Brecht left it ambiguous, but I cannot. I pick up my glasses and am sucked into a Proustian fruit wormhole, where I find myself in the company of other shortsighted pomophiles.

Largely hidden from the public eye, there exists a subculture of enthusiasts who have devoted their lives to the quest for fruit. With associations like the North American Fruit Explorers and the Rare Fruit Council International, the denizens of this fruit underworld are as special as the flora they pursue. The forest, from the Latin *floris,* meaning "outside," has always attracted outsiders. Since 1910, the word "fruit" has been used to denote an eccentric or unusual person. Writing this story meant getting to know fruit nuts, fruit smugglers, fruit explorers, fruit fetishists, fruit inventors, fruit cops, fruit robbers, fruitarians and even a fruit massager. These characters offer a glimpse into our planet's diversity—both botanical and human.

THE DEGREE TO which fruits can ensnare us is evident in Robert Palter's 2002 book, *The Duchess of Malfi's Apricots, and Other Literary Fruits*. A work of irrepressible mania, it's an 872-page anthology that attempts to catalog and discuss every occurrence of fruits in stories, songs, films, poems or other literary vehicles throughout history. One chapter even examines the conspicuous absence of fruits in certain books.

Palter, a retired professor in his eighties who had been part of the Manhattan Project that developed the atom bomb, is happiest when discussing snippets of obscure poems about fruits invested with "cosmic significance," such as Anthony Hecht's "The Grapes," full of clustered planets and little bags of glassiness.

He even gets into a dissection of fisting while discussing William Dickey's poem "Plum," citing the lines "Forcing their unlubricated entry, / a fist in spring & the bruised membranes clench / & convulse."

Palter concludes that the poem "is always on the verge of shattering the clenched delicacies of the traditional villanelle form."

I found Palter's number in the phone book and gave him a call. He wasn't particularly interested in edible fruits. "I deplore the fruits that I do get," he sighed, speaking from his home in Connecticut. But then his mood brightened as he explained that he had recently tasted a fresh fig for the first time. "I thought, 'there can't be such an organism—it's just too much!'" he exclaimed, excitedly sending short, staccato bursts of air into the receiver.

I asked him how he came to be so interested in literary fruits. "There's an obvious association with fruits and human life and love and sex and enjoyment," he told me. "But fruit rots! So there are negative connotations, as well. Fruits can represent political corruption. I can find you literary examples that make use of fruit to present any human feeling, sometimes rather subtle feelings—it's the whole spectrum."

Bob Dylan, in the liner notes to "Highway 61 Revisited," wrote ironically of someone "writing a book about the true meaning of a pear." Palter's book, which doesn't include Dylan's quote, notes that pears can have infinite meanings, whether as sexual objects, images of failed hopes, or metaphors of entropy. The essence of Palter's investigation is that the true power of fruits lies in their ability to seduce us.

He began by envisaging a short essay on fruits, but his research started accumulating at an alarming rate. He was soon overwhelmed by examples. "Every time I'd find another instance of fruits in a story, I'd say, 'Wow! I can't believe this!'" Librarians started referring to him as "the fruit guy" behind his back. After he'd gathered a mountain of fruit anecdotes, he approached the University of South Carolina Press about doing a three-hundred-page book. By the time he submitted his manuscript, it had swelled to six hundred pages. As it was being prepared for publishing, he kept sending in new material, until he was finally told that enough's enough.

According to his introduction, the project is inherently open-ended, the book merely an "interim report." He decided to end the book with no punctuation, as a sign of its endlessness. Long after publication, he still couldn't stop finding fruit episodes. As he put it in a reminiscence called *My Big Fruit Book*: "Involuntarily, and even against my conscious intentions, I persist in scanning for fruit everything I encounter in the way of print and pictures."

Toward the end of our conversation, Palter said that he was considering donating all his fruit books to the library. "I have to stop with the fruits," he said, letting out another huge sigh. Still, months after we spoke, he'd send me e-mails with fruit anecdotes. One contained his "recent discovery" of a scene in Javier Marias's novel *All Souls*. Listing the page numbers and the context—a faculty dinner—Palter recounted that during dessert the Warden "'insisted on decorating the bosom of the Dean of York's wife with a necklace made out of mandarin segments.' What a picture! Best, Robert."

AS FASCINATING AS literary fruits may be, I also wanted to learn the stories behind the actual fruits we eat. Our supermarket staples can be traced back to particular people and places. The Hass avocado began when Pasadena postman Rudolph Hass was persuaded by his children to not cut down an odd seedling. He patented his avocados in 1935; today Hasses account for the overwhelming majority of avocados sold worldwide. Bing cherries are named after Oregon's Ah Bing, a nineteenth-century Manchurian. The clementine is a type of mandarin baptized by Father Clément Rodier at his Algerian orphanage in 1902. The tangerine is a mandarin variant from the city of Tangiers. Dingaan's apple is named after an African chief who was murdered after murdering his brother. The McIntosh apple began with a broken heart.

John McIntosh was born in 1777 in New York. As a young man, he fell in love with Dolly Irwin, whose United Empire Loyalist parents were against the American Revolution—and their daughter's marriage. Eighteen-year-old McIntosh followed Dolly soon after she and her family fled to Canada. Tragically, by the time he arrived at their encampment in Cornwall, she had passed away. In a state of grief and disbelief, he exhumed Dolly's remains to confirm that she was really dead. After sobbing upon her decomposing corpse, he set out on foot, eventually settling on a plot of land near the village of Iroquois, Ontario. The terrain was overgrown with weeds, brambles, and bushes. Clearing it, he found twenty small apple trees, all of which soon died, except one—which bore exceptional fruit. Cuttings were grafted onto other trees, and by the early twentieth century, McIntosh apples were widely available.

Today, there are more than twenty thousand *named* varieties of apples—not including all the countless wild weirdos that never merited a moniker. So many apples exist that we can't even count them all. Forget an apple a day—you could eat a different apple every day for the rest of your life, or at least for the next fifty-five years. Some of the best taste like raspberries, fennel, pineapple, cinnamon, watermelon, broccoli, crayons or banana-hazelnut ice cream. There is a variety of rectangular yellow apple whose hollow core seeps mellifluous liquid as you eat it. It's like the natural version of that gusher bubblegum with juice in the center. There are black-skinned Gilliflower apples, ivory hued White Transparents, orange-fleshed Apricot apples, and others with deep red interiors. A few summers ago, in the heirloom apple orchard at Vancouver's Strathcona park, I came across a nacreous apple identified by its tag as a Pink Pearl. As I cut out the first neat slice for some friends, we all gasped: its flesh was bright pink.

Every time we eat a fruit, we're tasting forgotten histories. Emperors, tsars, kings and queens used to prize fruits. In one poem by Ibn Sara, oranges appear as damsel's cheeks, red-hot coals, tears reddened by the torments of love and balls of carnelian in branches of topaz. They've made us run riot metaphorically and literally. When pineapples first arrived in Britain they created a frenzy among aristocrats. Bananas once caused such a sensation in the United States that they were served as the pièce de résistance at the hundredth-anniversary celebration of the Declaration of Independence. They symbolized freedom, as they did in East Berlin when the wall came down—garbage cans were surrounded by banana peels, as they were the first thing the Ossis bought.

In the ongoing dispute between India and Pakistan over Kashmir, there have been moments of truce over *sharbat,* a local fruit juice. In 2000, some twenty-five thousand Indians offered Pakistani border guards *sharbat* as a gesture of goodwill. Both Israelis and Palestinians see the cactus pear as a symbol of their people. For Israelis, it represents their own prickly exteriors and sweet interiors. Palestinians see it as a symbol of patience, the patience it takes to peel and prepare the fruit, as well as the patience needed to cope with their ongoing challenges.

Fruits have fueled wars, dictatorships and the discovery of new worlds. A horse may have ended the Trojan War, but it began when Paris gave Aphrodite the apple of discord. Attica's figs spurred King Xerxes to embark on the Greco-Persian Wars. The Third Punic war was launched when Cato held up a fresh ripe fig and said, "Know this, it was picked two days ago in Carthage; that's how near the enemy are to our walls!" Albion brandishing oranges galvanized the troops of Lombard to invade Italy. The addictive fruit of the poppy plant was at the heart of Britain's opium wars with China. The nineteenth-century Maoris, who eradicated the Moriori in the Chatham Islands, went there because they heard it was the land of the karaka berry. In Scandinavia, battles that erupt between cloud-berry harvesters in Finland, Sweden and Norway have caused foreign affairs ministries to set up departments for "cloudberry diplomacy."

Fruits aren't what they seem. Red hearts and black eyes, capsules of sunlight and crystal drops of blood, as tempting—and deceitful—as the knowledge of good and evil, these sweet mirages have filled us with won-der from the start of time.

The earliest humans moved from tree to tree, eating their fill. Settling into sedentary agriculture, their descendents worshipped fruits. Reli-gions deified them, royalty clamored for them, poets seeking to express the ineffable carved symbols out of them, and mystics used them in ritu-als to facilitate visions. Fruits have activated our basest genetic instincts and elevated us to rapturous heights.

After all, Adam and Eve chose the fleeting taste of a fruit over eternal paradise. Buddha attained enlightenment under a fig tree. Mohammed said, regarding those who enter paradise, "for them there is a known pro-vision: fruits." The Aztec upper world of Tlalócan is an orchard full of glowing fruit trees. Tribes in the Malay Peninsula, such as the Jakun or the Semang, speak of a "Fruit Island" where dead souls end up. The Egyptian promised land was called Yaa, and according to a hieroglyphic epistle by Sinuhe, its trees were laden with figs, grapes and many other fruits. In Asgard, the Norse otherworld, apples keep gods young and immortal. Ancient Greece's Elysian Fields, reported Robert Graves, are apple orchards where only the souls of heroes are granted access. Avalon, meaning "Apple-land" or "Isle of Apples," is where King Arthur went to live forever and eat apples. In Jewish lore, when you enter paradise, you get eight myrtle berries and a standing ovation. Africa's Brakna nomads believe that heaven is full of gourd-sized berries. The Wampanoags of New England follow the smell of strawberries to get to the Spirit World. Thomas Campion described paradise as a place wherein all pleasant fruits do flow. Paramahansa Yogananda notes that "a Hindu's heaven without mangos is inconceivable!"

In Chinese mythology, peaches tended by Hsi Wang Mu, the Queen Mother of the West, are said to grant eternal life. She dwells in a palace encircled by a golden wall on the mountains at the summit of the world. Her garden is full of perfumed blossoms and trees dripping with blue-green fairy jewels. Beside a lake of gems, where invisible instruments play gentle melodies, Hsi Wang Mu's beautiful daughters serve peaches that take three thousand years to ripen. These fruits render the eater immortal.

In recent years, a scholarly debate has raged around literal interpretations of certain Islamic precepts. German linguist Christoph Luxenberg claims that today's version of the Koran has been mistranslated from the original text, and that the seventy-two virgins (*houris*) promised to martyrs are in fact "white raisins" and "juicy fruits." Luxenberg's chief hypothesis is that the original language of the Koran was not Arabic but something closer to Aramaic. Through an analysis of the paradise passage in Aramaic, he confirms that the mysterious *houris* become fruit—much more common components of the heaven myth.

The term paradise itself derives from the Persian language of Avestan. *Pairidaeza* initially meant an irrigated pleasure garden full of fruit trees. In the Islamic tradition, gardens replicate heaven. The same applied in ancient China, where the Imperial gardens were magical diagrams of the otherworld. Britain's Stephen Switzer, author of 1724's *The Practical Fruit Gardener,* wrote that "a well-contriv'd Fruit-Garden is an Epitome of Paradise itself." The glowing lights on Christmas trees are linked to the pagan Germanic belief in wish-fulfilling trees laden with numinous fruits. References to the divine crop up repeatedly in the scientific names for fruits. Taxonomists named cacao fruits *Theobroma,* Greek for "food of gods." A Latin binomial for bananas was *Musa paradisica,* or fruits of paradise. In 1830, grapefruits became *Citrus paradisi.* Persimmons are *Diospyros*: "fruit of the gods."

IN THE PAST, fruits were elusive treasures. When my father was a student in Eastern Europe in the 1950s, his elementary school's top prize for the year's best student was half an orange. Mark Twain considered watermelons chief of this world's luxuries. Thoreau saw apples as "fairy food, too beautiful to eat." In the early Middle Ages, fruits were so rare they were likened to angels' tears of joy. If you were able to taste a plum only once or twice in your life, you'd probably think of it as a holy mauve sphere speckled with golden pixie dust too.

Nowadays, fruits have become part of the daily grind. We have unlimited access: they're sold year-round, they're cheap, and they shrivel into moldy lumps on our countertops. Eating one is practically a chore. Many people even dislike fruits. Perhaps that's because, on average, fruits are eaten two to three weeks after being picked.

Our global economy demands standardized products: dependable, consistent and uniform. Having commodified nature, we're eating the shrapnel of a worldwide homogeneity bomb. I've purchased identical apples in Borneo, Brazil, Budapest and Boston. Many of the fruits we eat were developed to ship well and spend ten days under the withering glow of fluorescent supermarket lights. The result is Stepford Fruits: gorgeous replicants that look perfect, feel like silicon implants and taste like tennis balls, mothballs or mealy, juiceless cotton wads.

Real fruits are delicate, living things that need to be handled with care. Despite our manipulations, fruits are at their core rebellious and unpredictable. Apples from the same tree have different flavors. The time of day a fruit gets picked affects its quality. Individual sections of a single orange vary in sugar levels. Next time you find a good peach, try biting the bottom—it's sweeter. Fruits are ephemeral, meant to be enjoyed at the time of harvest. We've found ways of circumventing the restrictions of seasonality—distribution cold chains, precision agriculture, genetic meddling—but in doing so, we sacrifice flavor. Fruits today are as bland as they are ubiquitous.

This Faustian transaction includes unpleasant side effects. Chemical and pesticidal residues. Waxes and dyes. Uncontrollable oil hemorrhages. Banana republics. Irradiation and fumigation facilities. Cold-storage rooms where fruits rust for months on end. Billionaire fruit barons importing illicit substances on eighteen-wheelers from Colombia. Indentured laborers dying in tropical fields.

But things are changing. There are alternatives to monoculture, whether it's velvety peaches that gush nectar or heirloom pears dating back to the Renaissance. Growing these blasts of taste-bud bliss requires obstinacy, perseverance, and above all, passion. Fortunately, small producers' devotion is being mirrored by consumers and chefs, with the ensuing coverage in food media contributing to the emergence of rock-star farmers. The next step may be a rediscovery of lost tropical crops, many of which could reduce global hunger. In his book *Biophilia,* Wilson suggests three "star species," all fruits, that represent this hope: the winged bean, the wax gourd and the Babussa palm. We have much to look forward to.

For now, the fruits most likely to be found in fruit bowls the world over, based on the United Nations' statistics, are bananas (and plantains), apples, citrus fruits, grapes, mangoes, melons, coconuts and pears. Peaches, plums, dates and pineapples occupy a smaller dish off to the side. The developed world's bowl is bigger, and also full of strawberries. The developing world cradles infinite varieties of underutilized tropical fruits.

By discovering fruits, whether in our backyards or abroad, we can reconnect with nature, the realm of the sublime. To experience biophilia is to love a diversity that, as limitless as it is fragile, both haunts us and fills us with hope. This then, is the story of fruits, and of the intense connections between fruits and humans. A caveat: they can become an all-consuming preoccupation.

"Near complete absorption in an eccentric interest was seen as admirable and virtuous in the 19th century," explains historian of science Lorraine Daston, noting, however, that by the twentieth century, the single-minded pursuit of natural knowledge came to be perceived as "almost pathological, a magnificent but dangerous obsession." Delight has a price. Daston gives examples of how those intoxicated by the marvels of nature, particularly those with an unremitting attention to a single subject, face nervous collapse, isolation and, curiously, the same obstacles faced by any addict. The need to know has been perilous since Genesis.

I started dreaming of fruits almost nightly. I dreamed of finding important scrolls hidden under peaches at the grocery store. I learned how to play music with mangoes and take photos with oranges. I came across edible kaleidoscopes while hunting treasure inside labyrinth caves on secret islands. I dreamt of self-immolation, and in the heart of the flames, fruits appeared to me.

From the beginning, it was unclear whether I was hunting fruits or whether they were hunting me. It was in Brazil that I was first lured by the siren song of fruits, but the Newton-like epiphany that I had to tell their story came a couple of years later. I was sprawled on a poolside deck chair at Hollywood's Highland Gardens Hotel, reading about how fairy tales are the basis for our modern stories. Just as I began a paragraph describing the magic seeds that guide the hero through treacherous trials, a golden speck landed on the page. I looked up at the branch overhead. A moment later, a second speck fell onto the open book, right on the y in eternity. Pressing my finger into it, I picked up the plant particle to give it a closer examination. Slightly smaller than a peppercorn, it was oblong and covered in tiny bristles. I popped it with my pencil tip, and a tender yellow seed rolled out. My botanical knowledge was so rudimentary I wasn't even sure what part of the plant it might be, but my suspicions proved true. Fruits, so ordinary they're extraordinary, were summoning me.

Part 1

Nature

Ficus carica, fig tree

1

Wild, Ripe and Juicy:
What Is a Fruit?

Homer: Lisa, would you like a doughnut?
Lisa: No thanks. Do you have any fruit?
Homer: This has purple in it. Purple is a fruit.

—*The Simpsons*

W HEN I THINK about opening a book in front of the fire-place, I don't imagine myself reading about fruits," a family friend confessed when he heard about my research. But as the discussion turned to the fruits we loved as kids, he suddenly remembered how his girlfriend had, a few years earlier, surprised him by hiding a guava in the house. "I could smell it as soon as I came in," he said. He hadn't tasted one since leaving Israel decades earlier as a fifteen-year-old. When he found the guava, he triumphantly carried it into bed and curled up with it. After deeply inhaling its fragrance, he started kissing it like a long-lost lover and rubbing it all over himself.

"I made love to that guava," he groaned.

Fruit, inherently erotic, have a storied heritage as sexual accessories. In the medieval era, it was considered a turn-on for a woman to peel an apple and coddle it in her armpit until infused with her body odor, at which point she'd present the love apple to her lover. Plums and prunes were de rigueur in Elizabethan brothels. "Orange wenches" used to sell their bodies and fruits at theaters under the reign of Charles II. Fruits were aphrodisiacs around the world, whether loquats in China, gumi fruits in Persia or pomegranates in Tunisia. Brazilian tribes used to increase the size of their generative organs by tapping them with phallic, bananalike *aninga* fruits. The *Kama Sutra* gives instructions on performing fellatio in the manner of sucking a mango. Fats Domino got his thrills on blueberry hill. According to anthologist of erotica Gershon Legman, "sophisticates often insert into the vagina fruits such as strawberries or cherries" (those with false teeth or a dental plate, he added, should resist the temptation).

Figs have been used as sexual charms since Babylonian times. Today, at Montreal markets, hawkers cry out that each fig contains five grams of Viagra. Over coffee one morning, Montreal artist Billy Mavreas was telling some friends and me how he'd just returned from a trip to Greece, where he'd observed old men jostling each other to get at the first figs of the season. He could understand their ardor. "Whenever I open a fig," he explained, "I want to fuck it."

Some of us freak out when touching fruits. "Haptodysphoria" refers to the unusual, almost fearful, sensation that certain people feel when handling kiwis, peaches or other fuzzy-surfaced fruits. The word botanists use for this short, downy hair is "pubescence." The *Oxford Companion to Food* claims that "the peach, of all fruits, most closely approaches the quality of human flesh."

Others prefer soft, spongy melons. In Brazil, there is a popular proverb: "women for procreation, goats for necessity, young boys for fun and a melon for ecstasy." A blogger named monkeymask has posted about his melon mishap. Although the cantaloupe was at first too cold for his purposes, he microwaved it to bring it to a warmer temperature. He only realized that the insides were scalding when his privates sank into molten fruit magma.

The conflating of fruit appreciation with carnality began in the forest. Bonobos, sharing 98 percent of our DNA, are humankind's closest living primate relative. Japanese primatologists in the Democratic Republic of the Congo have documented rampant group sex when nomadic bonobos come across trees full of fruit. These sexual free-for-alls apparently allow bonobos to blow off steam and get along together.

Communal fruit sex was a fact of life in early agrarian societies. As with the bonobos, fruit harvests were a prime opportunity to unleash an orgy, wrote religious historian Mircea Eliade. These orgies had chaotic, holy underpinnings that Eliade described as "unbounded sexual frenzy," where anything could happen and all decency was dispensed with. Peru's annual *Acatay Mita* ritual involved men and boys getting buck naked, running amok in their orchards and violating any woman who crossed their path. Anthropologists have documented fruit-fueled group sex among tribes all over the world, including the Oraon of India and Bangladesh, the Leti and Sarmata people in islands west of New Guinea, the Baganda in Africa, the Fiji islanders and the Kana of Brazil. In Europe, harvest orgies lasted well into the Middle Ages, despite condemnations from the Council of Auxerre in 590.

Among the most enigmatic masterpieces of Western art is Hieronymus Bosch's *The Garden of Earthly Delights,* which features naked people frolicking with oversized berries. It seems to insinuate that the plant kingdom's strange eroticism serves a botanical function. Afterall, every time we eat a fruit, we engage in a reproductive act.

ALL FRUITS START as flowers. At their most basic level, flowers are the plant kingdom's sex machines. When, in the eighteenth century, it was discovered that flowers had male and female reproductive anatomy, the public and the church reacted with outrage. Botanist Carl Linnaeus's description of a flower as numerous women in bed with the same man was vilified as "loathsome harlotry."

Accordingly, the flower's vagina became called the "style," the penis became the "filament," the vulva became the "stigma," and the sperm became the "pollen." No matter what the name, flowers are a plant's genitalia. And whenever flowers bump uglies, they give birth to a fruit. Fruits are the result of a flower's ovule being fertilized. Fruits, therefore, are love children, offspring of a fragrant union.

Like that golden speck that fell in my lap in Hollywood, any plant part containing a seed is a fruit. In botanical parlance, a fruit is the developed ovary of a flower, alongside any other structures that ripen with it and form a unit with it. Fruits are basically plant eggs. In human terms, think of a pregnant woman: a fruit is the plant version of the amniotic bubble that contains the fetus. The baby is the seed (or seeds, as with quintuplets);

the entire spherical container in which the baby floats is the fruit. A fruit is how a plant gives birth.

Fruits are seed envelopes that contain within them the genetic coding that will further the entire plant. Their role is twofold: to protect and nourish the seeds, and also to facilitate the dispersal of the seeds. As we'll see, it's this second aspect that arouses libidinal forces.

I realize this discussion is somewhat more technical than the age-old dinner-table debate over whether tomatoes and avocados are fruits or vegetables. Sweetness usually plays a pivotal part in that dispute. The sweetness issue actually went to the United States Supreme Court in 1893. They ruled that tomatoes are vegetables because they aren't sweet. Rhubarb, which is actually a stem, was legally granted fruit status in 1947—because it's usually baked into sweet dishes.

Colloquially, then, and for agricultural tax purposes, a fruit has to be sweet, eaten at dessert, or at least clearly be a lemon. Scientifically, however, the definition is broader. Green peppers, avocados, cucumbers, zucchinis, pumpkins, eggplants and corn are all technically fruits because they contain the plant's seeds. Olives are fruits. Sesame seeds come from sesame fruits. Those luffa sponges we use in the shower are fruits of the *Luffa cylindrica* tree. Vanilla is the fruit of a type of orchid. Roses turn into rosehips. Lilies become beadlike fruits. Poppy seeds come from fruit pods whose sap is full of morphine. We bite into and spit out the husk of sunflower fruits to get at sunflower seeds.

My girlfriend, Liane, was dubious when I first informed her that flowers become fruits. "What about the fuchsias on our balcony?" she asked, pointing at hanging pots spilling out velvety blossoms. I predicted that they'd grow into some sort of seedpods once the flowers were fertilized by the bees buzzing around. Sure enough, a couple of months later, Liane noticed some dark maroon berries clustered amid the magenta petals. They had juicy, blueberry-like interiors. We bit into them hesitantly. They tasted insipid, like stray vegetation, but they were fruits nonetheless.

All pea pods, beans and legumes are fruit. A peanut is a fruit that grows underground. After pollination, peanut flowers burrow into the ground like frightened ostriches, allowing the fruits to ripen in the darkness. The top six crops in the world are all technically fruits: wheat, corn, rice, barley, sorghum and soy. Even though grains are small, they too contain seeds. Africa's savannas started producing grasslands that were full of wild cereals and grains around 14 million years ago. This allowed humans to descend from trees and evolve into bipeds.

A nut is defined as a hard one-seeded fruit. Many things called nuts really aren't. An almond isn't technically a nut; it's a seed found within a hard shell that forms inside a fruit that is actually a relative of peaches, apricots and plums. Yes, almonds grow on trees. Eating an almond is like eating the interior of a peach pit. A coconut isn't a nut at all; it's a fruit.

As with all attempts at categorizing nature, exceptions abound. Those seed-containing dots on the exterior of strawberries are called achenes; each of these achenes is actually a fruit. The red flesh is merely a structure that evolved to help disseminate the fruit seeds. A fig is a pod containing many tiny fruits. A pineapple is an inflorescence that fuses many berry-like fruitlets into a thorn-tipped aberration. So what's a vegetable? Any part of a plant that we eat that does not contain seeds. It can be a root (carrot), tuber (potato), stem (asparagus), leaves (cabbage), leaf stalks (celery), and flower stalks and buds (cauliflower).

Broccoli is densely compacted flower heads that haven't yet opened. If allowed to keep growing, the tiny green heads open to reveal pretty yellow flowers that become seed-containing fruit pods. Capers, which are flower buds, turn into large, seedy, caper berries. Spinach is the leaf of a plant whose flowers yield a minuscule, often prickly, fruit capsule. We brew tea with leaves, but tea plants reproduce themselves with fruits. Marijuana is smoked in flower form, but the pollinated fruit seeds are the way to grow a new plant.

There are an estimated 240,000 to 500,000 different plant species that bear fruits. Perhaps 70,000 to 80,000 of these species are edible; most of our food comes from only twenty crops. Pomology is about the fruits we eat; carpology studies the fruits that all flowering plants bear, whether eaten or not. Botanical fruits are stored in a place called a "carpotheque" (which sounds like the place taxonomists go to let loose).

A number of foods that we call spices actually consist of, or derive from, dried fruits: pepper, cardamom, nutmeg, paprika, anise, caraway, allspice, cumin, fenugreek, cayenne, currants and juniper. Mace is the lacy arillus of the nutmeg fruit. Cloves are actually dried flowers, although when pollinated, they do develop purplish fruits.

Until recently, spices were items of utmost luxury. Islands were ravaged and populations decimated for their spice bounty. In Europe, whole nutmegs and peppercorns served on golden platters used to be eaten straight up as dessert. As gifts, spices were comparable to the Patek Philippe watches or Swarovski swans given today.

The discovery of the New World itself was precipitated by a desire to find a faster route to the Orient for a dried berry that was worth its weight in gold. Pepper was so valuable that individual grains were used as currency. Lodgers used to pay their rent in pepper. Rome staved off invasion by paying Alaric the Visigoth and Attila the Hun annual ransoms of pepper.

Cinnamon is the bark of a tree, but when explorers found South America, they thought it was a fruit. The historian Garcilaso de la Vega wrote of "bunches of small fruit growing in husks like acorns. And though the tree and its leaves, roots, and bark all smell and taste of cinnamon, these husks are the true spice." Under the command of Gonzalo Pizarro, two thousand soldiers ventured into the jungles of Ecuador, searching for these fruit treasures. After over two years lost in the forest, only eighty naked, hysterical stragglers made it back to Quito—without any cinnamon fruits.

SPECIATION, or the emergence of new life-forms, thrives in isolation. Over time, species separated geographically end up evolving into novel forms. Hundreds of millions of years ago, there were two supercontinents, Laurasia (North America, Europe and Asia) and Gondwana (everything else). Many fruits originated before these landmasses started fracturing and drifting apart, which is why certain wild fruits can be found in multiple parts of the world. For example, it's been proposed that apples first appeared on Laurasia. After the supercontinent broke apart, the fruit evolved differently ways based on where it ended up. In North America, the Proto-Apples evolved into crab apples; in Central Asia, bears spent millennia grazing on increasingly larger and sweeter fruits, leading to the domesticated apple we enjoy today.

In 1882, Swiss phytogeographist Alphonse Pyrame de Candolle published *Origin of Cultivated Plants,* revealing for the first time the distinct places of origin for many fruits. Using a multichanneled approach combining linguistics, philology, paleontology, archaeology and ethnobiology, de Candolle was able to ascertain where the greatest variation in a species was found.

Peaches, apricots, cherries and plums all came from Central Asia; the banana and the mango had ancestral links to India; pears originated in the Transcaucasus; wild grapes first grew east of the Black Sea; untamed quinces and mulberries cavorted near the Caspian; the watermelon emerged in tropical Africa. In some cases, pinpointing the spot was clear: citrus had more varieties and parasites in South China than anywhere else, so its home was not disputed. In other cases, such as coconuts or dates, it proved impossible to find a fruit's birth spot. The center of origin might be too diffuse, or covered by a city, or perhaps the forest was destroyed for lumber. Eons-old changes in climate may also have eradicated many clues.

Going back even further, researchers have been probing the fossil record for evidence of the prototypical fruit. Life began billions of years ago with single-cell organisms underwater. Some of the earliest plants to emerge from the depths some 450 million years ago were ferns and mosses. And then, about 130 million years ago, the first fruits and flowers turned up in or near water. Darwin, whose theory of evolution struggled to explain the suddenness of their appearance and the speed with which they overtook the planet, called their rise an "abominable mystery." The playwright Edward Albee has characterized this moment as "that heartbreaking second when it all got together: the sugars and the acids and the ultraviolets, and the next thing you knew there were tangerines and string quartets."

Scientists believe that water lilies, whose fruits germinate underwater, and star anise, with its woody pentagram fruits, are among the world's oldest. The Annonaceae family, containing cherimoyas, soursops and custard apples, is another primordial cluster. The recent discovery of a 125-million-year-old fossil of an aquatic plant called *Archaefructus sinensis* ("the old fruit of China") has led to suggestions that it is the first fruit in history.

Plants started to produce fruit in order to disperse their seeds. One hundred million years before the emergence of fruit-bearing flowering plants (called "angiosperms"), there were other plants called gymnosperms. Conifers and cycads, like all gymnosperms, have seeds that aren't fully enclosed in fruits. *Gymno-* means naked; *sperm* means seed. Think of a pinecone: the seeds are contained in woody bracts, but the bracts are slightly open. A gymnosperm's seeds drop near the base of the parent tree whereas an angiosperm's enclosed seeds can be distributed much farther. The earliest forms of plant life didn't even have flowers. They oozed around, shedding spores into the sludge.

After dinosaurs became extinct 65 million years ago, mammals and birds rose to prominence. Fruits were there to feed them, and in turn, be disseminated. Soon enough, the world was covered by plants producing a multicolored array of seed boxes. The angiosperms thrived because they devised ways to have their offspring carried far away. By 45 million years ago, rain forests covered much of the globe. Fossilized remains of tropical fruits have been found everywhere from London to Anchorage.

As SEED DISPERSAL mechanisms, fruits have myriad methods of roaming the world. Certain fruits cast their fate to the winds. Seeds with parachutes, helicopter propellers or featherdown attachments can be carried miles away from the parent plant on currents of air. Think of a summer day, with dandelions, cottonwoods and milkweeds wafting along a gentle breeze. One liana fruit from Borneo is a long-distance glider that floats through the air, pushed along by upward air currents and equatorial zephyrs, until it settles, with its seed, far away from the mother tree.

Other fruits can swim. They float like lifeboats across bodies of water and sprout on distant shores. The coconut, which bobs for months along oceanic currents, is an example of aquatic dispersal. The palm fronds flirting on every brochure for white-sand beach resorts testify to its success.

The burrs that stick to clothes after a hike in the countryside are also fruits. Burdocks, cockleburs and sticktights intentionally attach themselves to the fur of any animal who brushes up against the plant, the idea being that by the time the clinger falls off, the seeds will have traveled quite a distance. Devil's claws are fruits shaped like vicious clamps designed to latch on to the hoofs of passing mammals. One of the most extreme examples of a hitchhiker fruit is the Sumatran bird-catching tree. Its fruits are covered with tiny barbed hooks and a sticky gum that glues itself to birds' feathers. Certain birds carry the fruit to other islands; less fortunate ones get their wings jammed by it, and they end up dying at the tree base, becoming fertilizer.

Other fruits scatter their seeds by bursting open with a dramatic explosion. Touch-me-knots and Virginia knotwoods open elastically and hurl their seeds into the air. Witch hazel fruits are like small AK-47s that fire the seeds yards away. Many decorative flowers, like violets or impatiens, morph into ballistic fruits. When a fruit pod pops off the squirting cucumber, seeds ejaculate through the air with rocket propulsion.

Sesame fruits pop open at maturity; hence the command "Open Sesame." The *Clusia grandiflora* is a kind of dangling claw fruit that opens when ripe like the jaws of a mechanized "win-a-plush-toy" arcade game. Pistachio shells "laugh" themselves ajar when ready.

The botanist Loren Eiseley was woken up one night by an unidentified bang. It was, he wrote, "Not a small sound—not a creaking timber or a mouse's scurry—but a sharp, rending explosion as though an unwary foot had been put down upon a wineglass. I had come instantly out of sleep and lay tense, unbreathing. I listened for another step. There was none." After looking around, he noticed a small button on the floor. It was the seed of a wistaria fruit he had brought into his home earlier. "The wistaria pods had chosen midnight to explode and distribute their multiplying fund of life down the length of the room."

In the wild, wheat and barley shatter spontaneously, catapulting seeds. Early in the Neolithic revolution, we chose to cultivate nonexploding— or indehiscent—varieties. The ancestral versions of fruits like cardamoms, peas, lentils and flax dispersed themselves mechanically. Human intervention has taught them to keep their big mouths shut. Dehiscent pomegranates, once painted on the shields of ancient Persians, were the inspiration for explosive hand grenades (*grenade* means pomegranate in French).

On a sunny day in early July, I stepped onto my balcony and was greeted by a sky full of cottonwood seeds pirouetting slowly through the air. Each of these floating fluff balls might possibly grow into a tree. A clump of thousands had gathered in a cotton candy pool at the base of the stairs. As I was standing in the sunshine, it hit me that I was surrounded by millions of swirling fruits. Then one landed in my eyelash, hoping to take root.

BEING EATEN is how many plants distribute themselves. To this end, they use color and sweetness the way certain European restaurants use unctuous pitchmen to lure in hapless tourists. The fruit sacrifices itself for the seed, hoping it will take hold somewhere far away. Genetically, plants want the same thing as any other species: survival and replication. Anyone eating a fruit helps it achieve its goal of making as many copies of itself as possible.

When birds eat small fruits, they eat the seeds as well. After passing through the birds' digestive system, the seeds are dispersed from above (trees grown from aerial droppings are known horticulturally as "craplings"). When squirrels bury acorns, a number of forgotten nuts will become new trees. Crabs eat coconuts and tropical almonds. Many fish eat fruits. When not devouring entire cows, piranhas like guavas, berries and the fruits of the *Piranhea trifoliata* tree. There are even tiny fruits dispersed by ants and other minuscule insects.

Some plants are so cunning that they developed fruits that resemble insects—in order to be eaten by that insect's predators. The fruits of *Scorpiurus subvillosa* look like centipedes. Other fruits imitate worms, spiders and even horned beetles. Birds carry them off, thinking they've snatched a squirmalicious snack.

All species coevolve with other species. In certain cases, a plant's evolutionary partner may be extinct, yet their fruits have somehow lived on. More than fourteen thousand years ago, giant sloths, mastodons, mammoths, elephantine gomphotheres and Hummer-sized beavers roamed the Americas. These animals, known collectively as megafauna, ate fruits like the osage orange, a knobbly green fruit that lacks a twenty-first-century diner. The neotropical forests of Central and South America are full of bulbous fruits that aren't being distributed in any way. These are called "anachronisms" by Dan Janzen and Paul Martin, two scientists who hypothesize that such fruits are missing their Pleistocene partners. Even avocados, prickly pears and papayas used to be gulped down whole, seeds and all, by fridge-sized armadillos called glyptodonts.

In southern Nepal, the horned rhinoceros's main source of food is the *Trewia nudiflora* fruit. The rhino eats it, and then excretes the seeds in marshy lands where they can grow into new plants. With the Indian rhino on the verge of extinction today, it's possible that the plant will also become an anachronism. The dodo bird, before it was wiped out, is believed to have eaten tambalacoques, fruits of the dodo tree. Without its evolutionary partner, the Mauritian dodo tree was threatened with extinction in the 1970s. A tambalacoque seed apparently wouldn't grow into a new tree without being digested and having its hard exterior abraded by a dodo. Although aspects of this phenomenon have been disputed, botanists now seem to agree that the seed can be germinated by passing it through a turkey's gastrointestinal network, which has led to renewed dodo tree groves.

Almonds would be nothing without their pollinators: bumblebees.

Almonds are California's largest tree crop. Six hundred million pounds of the nuts become two billion dollars annually, more than double what the state generates with wine exports. Even so, there is a shortage of the nuts. Growers can't meet the demand, a situation compounded by Colony Collapse Disorder, the mysterious bee epidemic. Each spring, forty billion bees are brought into California, some flown in from Oceania on 747 planes. Foreign germs, parasites and other pathogens invariably commingle amid the 530,000 acres of almond trees stretching from Bakersfield to Red Bluff. Aggravating matters are evolving strains of a virus, rampant overuse of pesticides and some freaky new mites.

One of the tightest mutualisms known in nature is that between figs and wasps. Certain types of wasps spend most of their lives inside figs: they are born, grow up and mate inside the self-contained fig universe. When the female wasp has been impregnated by a male, she flies out, bearing pollen on her body. She then squeezes into another fig's tiny orifice, losing her wings in the process, and pollinates the flowers. After giving birth to her spawn, she dies.

At the Greek harvest ceremony called Thargelia, men and women used to beat their genitalia with wild fig branches thinking that would help fertilize fig trees. When crops were meager, humans wearing garlands of figs would be burned alive on fig-wood pyres to ensure abundant harvests. If only they knew they needed wasps. Unbeknownst to most of us today, some of the figs that we eat, such as the Calimyrna or the Smyrna, may contain wasp remains, although these insect husks are usually broken down by an enzyme called ficin. Most of the figs that are grown commercially, however, are self-pollinating varieties. This means that the wasps aren't needed to transport the pollen because the flowers can develop fruits without being fertilized. The term for such a fruit is parthenocarpic, meaning virgin fruit.

In the past, all figs needed wasps. It wasn't until parthenocarpic figs materialized that humans took the reins. Plantings of self-pollinating figs dating back to 11,400 B.C. were recently uncovered in the Jordan Valley. Archaeobotanists now consider this to be the earliest evidence of agriculture, coming approximately a thousand years before the domestication of other early crops like wheat and barley.

Despite their branding, many seedless fruits actually contain small, sterile seeds. If you look closely into seedless grapes, you'll see little aborted embryos, minuscule enough to not interfere with the fruits' texture. Seedless watermelons invariably contain white seed ghosts. Some parthenocarpic fruits, such as stoneless mangoes, plums and avocados, can be pretty unusual.

But then parthenocarpy is itself an unusual bargain. The plant says, "Fine, I'll stop producing seeds so that you'll eat me happily but in return, you'll spread my DNA far and wide by propagating orchards full of my clones." Seedless fruits haven't therefore lost their *raisin d'être*. Seeds used to be the way plants made copies of themselves. Today, humans can help them multiply more efficiently, mainly because we oversee most arable land. Plants evolved parthenocarpy to ensure their continued existence. Or we selected them; whichever you prefer.

NASA's freeze-dried peaches marked fruits' entrée into space. Now astronauts are growing fresh strawberries on board flights to Mars. Chinese seed satellites are orbiting Earth in order to see whether cosmic radiation and microgravity will lead to increased fruit yields. In Canada, Tomatosphere is the name of a program teaching students to plant and observe tomatoes whose seeds have spent months in space shuttles. Fruits are along for the ride: stuck in truck tires, ballast water and loading containers, they accompany us wherever we go.

Just as we enjoy them for their taste and nutrients, fruits have successfully enlisted us to extend their reach. In this sense, we're being manipulated by fruits. Our payoff is a healthy snack; their payoff is countless hectares of orchards tended all over the galaxy. Next time you eat a raspberry, pay attention. The berries look and taste so delicious that you probably don't even notice that those seeds are passing through your digestive system intact, ready to return to the earth and blossom anew.

As my guava-groping friend knows, fruits are arousing. Many are named for their ribald aspects: sodom apples, tit of Venus peaches, women's breast apples and maiden's flesh pears. The udder-shaped nipple fruit, also known as the titty fruit, is an egg-sized orange freakazoid covered in nipplelike nobules. The conquistadores named the *vainilla* bean after the Latin word for vagina. Buttocks, nipples, bosoms, thighs and fingers have long been employed as names for different cultivars. (A cultivar is a *culti*vated *vari*ety that has desirable attributes that are distinct from other cultivars.)

Humans have always delighted in the similarities between fruits and our erogenous zones. An Egyptian papyrus from more than three thousand years ago equated pomegranates to breasts. Plums, peaches, cherries, apricots, have all had their exteriors compared to posteriors. The quest for the callipygian ideal seemed to find its grail in melons, whether it was Apollinaire comparing women's bottoms to melons grown in the midnight sun or James Joyce almost rapping in *Ulysses* about Leopold kissing the "plump mellow yellow smellow melons of her rump, on each plump melonous hemisphere, in their mellow yellow furrow, with obscure prolonged provocative melon-smellonous osculation."

Lest we think fruits are exclusively feminine, consider the banana. The candlestick salad, a popular dish in the 1950s, featured an upright banana shaft rocketing out of a pineapple ring crowned with molten whipped cream. "Fruits" were an aristocratic code word for sperm, as evinced by a French poem about two cousins going plum picking: "She brought back many fruits / but these fruits weren't plums." Avocados for the Aztecs, figs for Berber nomads, apples for Servius: all were called testicles. Mangosteens are said to resemble the interior of scrotums. Lychees are "naked balls," wrote Georges Bataille. A variety of fig from Naples called pope's testes has an almost transparent strawberry-pink flesh.

Fruits intentionally send out attraction signals. No wonder we go bananas for them; they've programmed us. Fruits reproduce themselves by making us want them.

Humans are willing to go all the way—perhaps without even realizing it—for fruits. Like us, they are alive: they pant, perspire, and pop open. They even possess a form of intelligence: bananas and oranges connected to lie-detecting polygraphs have been shown to respond to mathematics questions in experiments by Dr. Ken Hashimoto and Cleve Backster. Asked how much two plus two is, the plants emit a hum that forms into four peaks when translated into ink tracings. In recent years, molecular geneticists have deepened their understanding of plant perception. By decrypting plant signaling, we've learned that flora have the sensory capacity to compute everything from temperature to light, pulsating with electrical receptors when under threat and flooding areas under attack with toxins. Jeremy Narby, in *Intelligence in Nature,* explains how plant cells communicate information using RNA transcripts and protein links. In this way, "plants learn, remember, and decide, without brains." The Japanese have a word for the "knowingness" of the natural world: *chi-sei.*

It's clear that both sides have evolved the ability to influence the other. And what fruits want from us is the same thing we want from them: survival.

2

Hawaiian Ultraexotics

With an apple, I will astonish Paris.

—Paul Cézanne

KICKING THE FEBRUARY slush from their boots, friends are pouring into my apartment for a cocktail party. Montreal's minus forty-degree winters may inspire indie-pop odes, but they aren't exactly conducive to a thriving fruit culture. Still, wondering what might turn up, I asked each guest to bring a fruit they've never tasted before. But no one seems to have brought any fruits at all.

Then, shortly after midnight, the doorbell rings. I open the door, and my shivering friend Karl hands me a bright pink orb. The size of an ostrich egg, it's winged with flaming orange-green flaps and topped with a mane of dead flower petals. It's a dragon fruit, Karl says, from Vietnam. He just picked it up in Chinatown. It looks like an emissary from Mars.

An excited crowd watches as I cut it open, revealing crisp white flesh dotted with small black seeds, like a solidified Oreo milk shake. Karl hands out sections that, with their shocking pink rinds and black-and-white interiors, resemble slices of zebra meat. The texture is akin to watermelon, the seeds as inconspicuous as a kiwi's. The delicate flavor is vaguely reminiscent of strawberries and concord grapes. Some people say they find the taste too subtle, but there's something about the flavor's very restraint that perfectly complements the dazzling exterior.

Hunting for other fruits that can match the excitement of dragon fruits, I start making regular treks down to Chinatown. The more outrageous, the better. I find heart-shaped lychees that gush sugary nectar. Grape-sized longans, beneath their dusty beige peel, are filled with gelatinous interiors. Their sweet-tart bursts of juice taste gently spiced with nutmeg, cloves and cardamom. I love the way you can eat kumquats, like mini tangerines, peel and all. Pepinos are purple-flecked cucumber-melons that, unfortunately, look better than they taste. The same is true of the kiwano, a Day-Glo orange spiky lump that could be mistaken for a radioactive horned toad. Containing barely palatable slimy green seeds, its sole purpose appears to be visual appreciation.

There are a lot of options, which might explain why Quebeckers eat more fruits than other Canadians. The most flavorful fruit I come across, the mangosteen, isn't flashy at all. Known as the queen of fruits in Southeast Asia, its hard, purplish and ocher shell, crowned by a woody flower cap, is sliced around its circumference and then twisted open. The interior of the fruit is fitted with a half dozen ivory-white fruit sections that look like garlic cloves and taste refreshingly majestic. Each self-contained section is just firm enough to suspend the incomparable juice in a perfect degree of tension. I could say that it tastes like minty raspberry-apricot sorbet, but the only way to truly know a mangosteen is to try one. Philosophers have discussed the impossibility of conveying the flavor of a fruit to someone who has never tasted it. As David Hume put it: "we cannot form to ourselves a just idea of the taste of a pineapple, without having actually tasted it."

I start bringing mangosteens to parties. Some people are as impressed as I am. Others wonder if its name means it's somehow related to mangoes, or if the "-steen" makes it Jewish. I also start noticing that some people aren't at all interested in trying the fruit-queen. "I had one of these removed last year," a disgusted friend tells me. "From my back."

On a trip down to Manhattan, Liane and I pick up a bunch of Chinatown fruits as a gift for our friend Kurt Ossenfort, in whose guest room we always stay. An artist who used to tie paintbrushes to trees so that oaks could do paintings in the wind, Ossenfort always seems to be involved in glamorous "giglets" such as videotaping photo shoots for *Teen Vogue,* filming documentaries about Thai elephant orchestras or designing penthouse suites for lawyers representing the World Trade Center. We arrive at his Fifth Avenue apartment and present him with the mangosteens, dragon fruits, sapodillas, dukus and longans. He's appreciative, but also a tad concerned about how we got them over the border. Mangosteens and dragon fruit, he says, are illegal in the United States. I'm stunned—why can you get them in Montreal, but not in New York? It has something to do with bringing pests into America, explains Ossenfort. Eating them, we laugh off the inadvertent smuggling.

Ossenfort's own fruit bowl is full of red plums dappled with yellow splotches. Their little stickers say that they're pluots: hybrids of plums and apricots. They taste like the juiciest, sweetest plum imaginable, with a hint of apricot flavor. Tasting them conjures images, remembered or dreamed, of a dusty summer afternoon, somewhere in Croatia, of the first photograph I'd ever taken: it was of a fruit tree. Its fruits tasted something like these pluots.

I ask Ossenfort where on earth he got them. You can get pluots anywhere, he says. But he happened to know these ones were especially good because the Fruit Detective had tipped him off.

"The Fruit Detective?" I ask.

"His real name is David Karp," explains Ossenfort. "And he knows everything about fruits." Ossenfort shows me footage he filmed of the Fruit Detective jumping around in foliage and stalking pineapples. Karp was wearing a pith helmet and had a lazy eye. He used all sorts of esoteric paraphernalia, such as a refractometer to gauge the sweetness of fruits and a knife identical to those used by members of Shanghai's Green Gang to shiv enemies.

The Fruit Detective, I'm amazed to learn, is the son of an inconceivably wealthy copper magnate. Renowned in certain Upper East Side circles for having published translations of sixth-century Latin poets at the age of twenty as well as for getting near-perfect scores on his SATs while tripping on LSD, Karp's career as a high-powered stockbroker crumbled as he became addicted to heroin. At one point, he was flying regularly by Concorde from Paris to New York to score drugs. He even became a dealer. After going through recovery, he reinvented himself as a fruit connoisseur to impress a girl he was in love with. Although he didn't get the girl, he did become a fruit junky. He even calls eating fruits "a fruit fix."

He's a shining light. His compassion for unknown fruits, even if the affinity verges on the compulsive, is inspiring. I'm convinced he's the hero of a story not yet written.

BACK IN MONTREAL, I head for Chinatown and pick out the most extravagant dragon fruit in the city. It's a gift for an editor at Air Canada's in-flight magazine. Describing the pluots, and David Karp, and the bounty of fruit in Brazil, I pitch a story about exploring for fruits. She is blown away by the dragon fruit, and serves it to colleagues at their editorial meeting. A few days later, I get a call. They've commissioned a feature story asking me to follow the Fruit Detective on a fruit escapade.

When I call David Karp to tell him about the story, he informs me that he is going to be profiled by *The New Yorker*. One of their staff writers has asked to spend a week with him later that summer. He says I can only interview him after that piece appears.

Trying to keep the story alive, I ask if he'd join me on a trip if the magazine pays for it. That changes everything. If they send us to Alaska to find cloudberries, he'll gladly do the interview. He then tells me that he's already been written about in *Seventeen,* but that pubescent girls haven't yet started throwing their nighties at him. His favorite part in *Lolita,* he says, is when they cross the border into California and the agricultural officer asks if they have any honey. We end the conversation with him talking excitedly about cloudberry hunting. He sends me pages of notes on the fruit, a raspberry-like orange arctic fruit with a strong musky flavor that grows in places like North Pole, Alaska. According to his notes, cloudberries are often found in bogs full of insects that tear off chunks of skin with their "horrid mandibles."

Unfortunately, my editor squelches the Alaskan cloudberry dream. Instead, we decide to focus the story on fruit tourism, something I'm not even sure exists. My assignment is to find other people like the Fruit Detective and examine the trend of traveling for fruits.

I start making a list of fruit tourism destinations, such as Bologna's Garden of Lost Fruit, Yamanashi's postmodern glass-and-steel Fruit Museum, and an island in the Nile river called Gazirat al-Mauz ("Banana Island"), where visitors can sample myriad bananas. I tell my editor that it looks like the best place to go hunting for fruits is Malaysia. "We're not going to send you to *Malaysia,*" she says, rolling her eyes. They will, however, send me to Hawaii . . .

LEAVING THE Big Island airport, my taxi driver starts singing "Welcome to Hawaii" into a little microphone over a backing track of slack-keyed guitars. His voice crackles through an amplifier he's set up on the dashboard next to a hula doll in a grass skirt. As we skim along the winding road on the Kona coast, the sky and ocean merge into a blue infinity. Lush greenery bursts out of the black hardened lava. At a traffic light, a panoply of trumpet-shaped flowers perfumes the air. It's all growing out of what is essentially a volcanic mountain island that's still squirting out red, purple and golden streams of lava. We pull over briefly at my hotel, where I check in and snatch some of the complimentary papayas and mangoes at the buffet. They taste terrible.

Twenty minutes later, we pull up at a dusty side street next to a rusty macadamia nut processing factory. "Many *mahalos,*" sings the driver into his microphone. "This is it: Napo'opo'o Road."

I look around. I don't see anything besides trees and a dirt path. "Where's the market?" I ask. "It's down that way," the driver says, pointing at a sign that says KONA PACIFIC FARMER'S COOPERATIVE. Walking along the path, I notice a couple of vendors have set up goods on some picnic tables. From a distance, it looks less like a market, and more like a rummage sale. A few men are carving sculptures of tiki gods from blocks of wood. I tell them I'm looking for Ken Love, president of the Hawaii Tropical Fruit Growers' western chapter.

"Yo, Ken, you have a customer," shouts one of them.

A man pops his head out from under a picnic table and waves. I walk over and shake his hand. Ken Love is unshaven, bulky and sweating in the heat. His enormous Hawaiian shirt is caked in farm grime. He takes off his floppy green hat and wipes his brow, revealing a balding pate girdled with curly graying hair. His granny glasses are smudged and a carved pipe juts out from under a bushy mustache. Despite his huge smile, there's something kind of shifty and mischievous about him. I like him immediately.

His stall contains dozens of different types of fruit. Each of these is displayed alongside photos and text describing the fruit's characteristics. Ken Love calls them ultraexotics, to distinguish them from the exotic-yet-commonplace mangoes, papayas and pineapples of commerce. I taste an acerola, a bracingly tart red berry that, he says, has four thousand times more vitamin C than an orange. He shows me green thumb-shaped bilimbis, starfruit relatives that his wife, Maggie, uses to make chutneys. We slice open a dusky brown chico—it tastes like maple syrup pudding. The grape-sized wampees, he says, counteract overindulgence in lychees. In China, eating too many lychees is believed to cause nosebleeds—popping wampees apparently stops the bleeding. I eat bignays, gourkas, sapotes, mombins, langsats and jaboticabas of all shapes, sizes and colors. There are so many new fruits that I start losing track, their names blurring in my notepad. I feel like I've somehow ended up in Neverland.

The bounty of the region is what originally persuaded Love to move here. "I used to be a photographer in the Midwest, and I'd come out here for work. What drove me nuts was seeing all the papayas and mangoes rotting on the sides of the streets. 'Why aren't people doing something about them?' I wondered."

As he started digging around, he kept finding more and more strange fruits. Hawaii is at the crossroads of the East and West, and every wave of immigrant always arrived with seeds in their pant cuffs, in their pockets and sewn into their shirts. "They all wanted that golden nugget from back home," explains Love.

Some of these seed introductions, such as Himalayan berries and the banana poke (a type of passion fruit) have taken over the island, snuffing out local plants and poisoning animals. Hawaii is considered the world capital of invasive species. Charles Elton, in *The Ecology of Invasion by Animals and Plants,* describes Hawaii as "quite an exchange and bazaar for species, a scrambling together of forms from the continents and islands of the world."

It isn't only humans who've brought seeds to the islands. A number of them simply floated in on air currents, explains Alan Burdick in *Out of Eden.* Scientists sticking nets out of plane windows have cataloged thousands of airborne species drifting above Hawaii. Such arial plankton accounts for only 1.4 percent of the seeds that arrived before humans. According to the biologist Sherwin Carlquist, most seeds and fruits arrived either inside birds' digestive systems or stuck to their feathers and feet. The rest washed in with the ocean current, a situation compounded in recent years by the minute life forms arriving in the bilge water of ships.

Hawaii today is full of immigrant fruits growing buck wild. Love's mission is to catalog, promote and sell them. Most Hawaiians don't even realize what's growing in their backyards, he says. They'd rather eat subpar fruits that pass through the industrial food chain than the fresh fruits growing all around them. I think back to the unimpressive fruits at the hotel buffet and imagine them making the journey from a farm in South America or Asia all the way to Hawaii.

People aren't eating their local ultraexotics, Love says, because they simply don't realize they exist. His goal at the farmer's market isn't so much to sell fruits, it's to educate people about the vast array of fruits on their doorsteps. When a family arrives at the co-op, he starts telling them all about the different fruits. They purchase a bunch of tropical apricots, mountain apples and Surinam cherries. I ask two kids what they think of the market. "Our parents said that we had to do something educational today," says one tween fingering a bignay. "This is educational but it's not too bad. It's fun, actually."

Two cougars in pearl necklaces scamper about while paraphrasing Love's descriptions. "I learned something today," says the one in purple Yves St. Laurent heels, picking up a bunch of "heavenly" rambutans. Covered with hairy tendrils, somewhat like a sea urchin, rambutans contain a delicious white lycheelike interior. As Love's data sheet explains, the fruit's name is derived from "rambut," the Malay word for "hair." The women titter about the rambutans' resemblance to hairy testicles.

Tugging on his pipe, Love explains that fruits have a transformative power on people. "Have you ever seen a Russian guy taste a jaboticaba for the first time?" he asks, referring to a fruit that "looks like an alien embryo and tastes totally out of this world." A few weeks ago, some Vietnamese women began crying when they saw his otaheite gooseberries. "Their mom's otaheite tree in Saigon got chopped down when they were children, and they hadn't seen the fruit since they were little kids."

Throughout the day, I keep on tasting the different fruits, amazed at the diversity. At one point, Love hands me an oblong berry, the size of a pinkie tip. I pop it in my mouth and bite. A pleasant spurt of juice coats my tongue. Love instructs me to spit out the seed. He then hands me a lime and asks me to taste it. It's sensationally sweet. I ask him if it's some sort of sweet lime. No, he says, laughing. I'm merely experiencing the aftereffects of the miracle fruit. Through some quirk of biochemistry, this small red berry has a miraculous effect on the palate: it makes all acidic foods taste sweet. It coats taste buds in a liquid that, for approximately one hour, alters our perception of all sour foods. After eating a miracle fruit, pickles taste like honey. A bologna and mustard sandwich tastes like cake. Vinegar tastes like cream soda. It's nature's NutraSweet.

As I sit there in a blissed-out miracle-fruit trance, hours fly by. At closing time, I help Love pack up the ultraexotics. He offers to give me a tour around the island the next day.

That evening, in the hotel lobby, the miracle fruit is still making my soda water taste slightly sweet. As I'm going over my notes, the singer of the lounge band sits down next to me. She has just finished a flamboyant rendering of Roberta Flack's "Feel Like Making Love." Her name is Priscilla. She is a transsexual.

"So whatcha doing in Hawaii?" she asks, in a husky baritone.

"I'm a journalist," I answer. "I'm working on an article about exotic fruit."

"I'm an exotic fruit," she purrs. "Write about me!"

As she sashays back to the stage, I head up to my room. Walking down the hallway, with its green shag carpeting, green fern wallpaper, and green ceiling, I feel as though I'm hurtling headfirst through a green tunnel into some sort of magical plant realm.

THE FOLLOWING MORNING, Love picks me up in his pickup truck. He has brought a flat crate of yellow apricot-like Japanese fruits called loquats. "I'm quite loquacious about loquats," boasts Love, explaining that he has amassed more than five thousand pages of notes on the fruit. His fixation began in Japan, where a former lover reached up and pulled one out of a tree. "Loquats were forbidden to common people in China because of a legend about a carp swimming upstream who turned into a dragon by eating loquats. The emperor said, 'Hey, I don't want the common people to eat loquats and become strong like dragons and kick me out of office,' so they were banned."

Loquats are just the tip of his obsessiveness. Love is an Asiaphile who has been on many expeditions to the Far East. He once passed out in Singapore after eating sashimi of dog brain filet. He recently wrote an Internet guide to 1,530 Japanese restaurants in the United States. ("I actually went to three hundred of them.") He loves taking photographs of geometry, and, as we drive, he points out the Euclidian forms in the palm trees, in the steering wheel and in the fruits we're eating. His bylines in magazine pieces state that he "has been involved in agriculture since he was first spanked for tearing up the front yard to plant beans to bring to his new kindergarten teacher."

Our first stop is his farm. After driving up a boulder-strewn path, he shows me a coffee bush, covered in red berries. It never occurred to me that coffee came from a fruit.

Getting out of the car, Love explains that the hardened lava coating the island makes it exceedingly difficult to plant trees. To demonstrate, Love suggests we plant a lychee tree. The black volcanic soil is so hard that I need a pickax, not a shovel, to make a dent. Still, planting the tree is a wonderful experience—and also a prime fruit tourism activity. A lychee tree named Adam is growing on that farm at this very moment.

En route to our next destination, Love talks about the dozens of enthusiasts he knows who travel around the world looking for fruit. The idea of a community of amateurs trekking across the globe and into rain forests to find rare fruits fascinates me. We keep passing fruit trees lining the sides of the roads. "Here you don't stop to smell the roses, you stop to eat the fruit," says Love, as we jump out to check on some rare gru-michamas, sort of like Brazilian cherries, blossoming in the back of a Chevron gas station. With their slightly resinous undertones, they taste like cherry cola. In a nearby ditch, pineapple crowns jut out from spiky bushes. I somehow imagined that pineapples grow on trees, yet here they are bursting out of knee-high plants.

At lunchtime, we pick up Ken's plump, bespectacled sixteen-year-old daughter Jennifer and head to their favorite Chinese restaurant. We pore over a coffee table book by a Miami fruit hunter named William Whitman. It's full of odd photographs of the author holding monkeys and glowing red fruits. Love starts reminiscing about John Stermer, an eccentric fruit experimenter, who used to walk through his Hawaiian orchards in the nude. As he is raving about Asian fruits, Jennifer looks at me and says, "Welcome to the agony of my reality."

Our next destination is Lion's Gate B&B, a fruit-centric inn with an orchard full of pomelos, jaboticabas, Surinam cherries and rambutans. The owner talks about his "hot years" as a young lieutenant in Asia. "In Japanese folklore, the *tanuki* is the badger of excess. Sculptures depict him with money in one hand and sake in the other, going to town with a hard-on. I was called Tanuki when I was in Japan."

We then visit a vacant lot Love bills as a fruit tourism destination under construction. It belongs to Carey Lindenbaum, a sinewy Californian lawyer who moved here to grow fruits. "It'll be a B&B with all sorts of organic tropical fruit that you can pick yourself," says Lindenbaum, pointing at a weedy, rocky patch of land. Nestled between the limbs of a nearby tree is her home, a small wooden tree house. As she describes the future orchard, her pet donkey keeps pushing me away with its nose. "She's really jealous and possessive," says Lindenbaum. "She likes to invade personal space."

The next stop is George Schattauer's private orchard. Love is the custodian of Schattauer's rare trees. In exchange for tending the coppice, Love can sell excess fruits at his farmer's market. It's a beautiful garden, filled with wonders such as the egg fruit, a golden, mango-sized fruit shaped like a teardrop. Near the entrance is the noni: a lumpy, gnomish fruit that smells like dirty socks. It's also called the vomit fruit, says Love, but some people think it cures cancer. It can't be eaten raw because it's too tough; but in juice form, it became the locus of a health craze that swept through the diet-obsessed 1990s.

A dog follows us through the orchard, eating fallen mangoes. At the house, Schattauer points out a large orange tree. It's the first orange tree in Hawaii, he says, brought here by Captain Vancouver from Valencia in the eighteenth century. Before leaving, we walk under the drooping branches of a jackfruit tree. Near the trunk, there's a cavelike grotto formed by the surrounding foliage. Under the branches, a number of massive jackfruits shine like carbuncles in the darkness. (One of Schattauer's fruits, weighing more than seventy-six pounds, won the Guinness world record for biggest jackfruit.) Love squeezes them to test for ripeness, and then uses garden shears to cut off one the size of a fat six-year-old. It looks like one of those alien pods in *Cocoon*. He hands it to me. It is exuding a sticky, milky-white goo.

We sit down in the driveway and open it up, unleashing its sulfuric essences. It stinks. Naturalist Gerald Durrell describes its fearsome aroma as "a cross between an open grave and a sewer." He exaggerates, but only slightly. Inside are amber, honey-drenched segments. I'm mesmerized, but also scared. The jackfruit's stench is primal, uncomfortably animalistic. It's nature: unshaven, bulky, and secreting discharge. Love moans with pleasure as he eats it, wiping his nectar-sodden hands on his pants. I take a timid bite. Love offers me another sopping handful. I can't eat anymore, I explain. Love shrugs, taking another mouthful. I want to like the fruit as much as he does, but its aroma is too penetrating, too terrifying. As Love wraps up the fibrous parts to make jackfruit jerky, I feel disappointed in myself. My palate isn't yet ready for the gustatory delights of the exotic fruit kingdom.

As the sun sets, we head to a pizzeria to meet Kent Fleming, a tall professor at the University of Hawaii and the author of *Agtourism Comes of Age in Hawaii*. "There's tourism for everything," he says, citing disaster tourism and food tourism. Fruit tourism, according to Fleming, consists of the sort of farm and agricultural ("ag") visits I'm experiencing on this trip. "Agtourism is an *ed*-venture," he quips. Alongside creating an awareness of the countless varieties of fruit available, fruit tourism is a hybrid of educational and adventurous ecotourism that is concerned with developing models of sustainability in rural areas. It's a way for traditional family farms to create a new market in the face of corporate farming and for newfound fruit visionaries—like the people I've been meeting—to create viable business opportunities.

He explains that he's hoping to convert the orchard at the nearby university into a model fruit tourism destination. "It will be a central resource for travelers and farmers. We'll have a shop where you can buy fruits and books or find out about fruit tourism. You'll be able to have a local coffee and white pineapples. We'll have a hundred types of weird fruit and tons of different types of avocado—but not the Hass avocados you get on the mainland. We don't even let the pigs eat those." Hawaii, he says, is the one place where you can travel to taste the world's most exotic fruits without facing exotic dangers such as having insects lay eggs in your back that hatch into squirming worms. After a few beers, Fleming starts telling ribald jokes. At one point, he asks me if I've ever heard of a Father Nelson. He then puts me in a full-Nelson headlock and starts humping me from behind.

AFTER SAYING GOOD-BYE to Love, I board a plane to Hilo, on the other side of the Big Island, the following morning. The woman sitting next to me starts talking about "emissaries of light" while making a figure eight with her fingers. She explains how she made an imaginary contract between her and all the citizens of planet Earth, wrote "void" on it in red ink, shrunk it to the size of a postage stamp, and then burned it in a violet flame as the ashes dissipated into the light above.

My first destination is Onomea Orchards, a tropical fruit farm belonging to Richard Johnson, a retired manager from Intel. Confident and businesslike, he is growing rambutans, mangosteens and durians for profit. He's convinced that these fruits will soon be as popular as kiwis. Although they haven't been allowed into the United States in the past, he says that Hawaii has invested in irradiation facilities that are now allowing ultraexotics to be exported to the mainland.

He shows me some durian blossoms, which are hermaphroditic (meaning the flowers contain male and female parts together). They require hand pollination. A fine mist filters down through the rambutan trees. "Cicada droppings," he says. I tell him that I'm curious about people traveling for fruits. He, like Love, is familiar with a number of fruit hunters. He tells me that Ken Love and some other Hawaiian fruit fanatics refer to themselves as "the Hawaiian Mafia."

"Have you heard of the fruitarians?" Johnson asks.

"Fruitarians?"

"You know, people who eat exclusively fruit. There's a lot of them over in Puna [a nearby town]. We call them Punatics." He suggests that I speak about fruitarianism with Oscar Jaitt, who lives nearby. Jaitt, he explains, has traveled all over the tropics looking for fruits and is the owner of www.fruitlovers.com, a website that sells exotic fruit seeds. He also produces a line of fruit lotions called Alohatherapy ("made with aloha").

Arriving at Oscar Jaitt's address, I walk through his serene Buddhist garden and knock on the door of his hexagonal wood cabin. A very mellow man in a white beard appears, wearing billowing purple Zubaz pants that taper at the ankles. "Fruit farming is spiritually fulfilling," he says, as we stroll over to his fruit orchard in the adjacent lot. "It allows you to see the miracle of the cycles of life."

I ask him if he's a fruitarian. He laughs, and says that he isn't quite a fruitarian, although he mainly eats fruits because he's a raw foodist. He does confirm the existence of fruitarians, many of whom indeed live in Puna. "Fruitarians have no constipation," he says. "They laugh at ex-lax."

We taste some jaboticabas. They look like oversized deep purple grapes, matching his pants. Because the fruit grows directly on the tree trunk like some sort of sweet fungus, the best way to eat one is a "jaboticaba kiss." In Brazil, kids sneak into other people's backyards and kiss them off the trees. Brazilian writer Monteiro Lobato describes the sound of a jaboticaba kiss as "plock, pluff, pituy."

Jaitt shows me a rollinia tree whose head-sized fruits taste like lemon meringue pie. Nearby are two small trees he's especially excited about: the peanut butter fruit and the blackberry-jam fruit. "The peanut butter fruit looks like a red olive," he says, "but tastes like Skippy and even has the same texture." The accurately named blackberry-jam fruit has a yellow exterior and a black interior. Jaitt says he knows some people in Honolulu who've been serving peanut butter fruit with blackberry-jam fruit and breadfruit. Kids apparently go crazy for these all-fruit PBJ sandwiches.

Looking up, I notice a long green fruit dangling from a branch. "Oh, hey, look at that," says Jaitt. Walking over to get a long pair of clippers, he snips it off for me. It's an ice cream bean, also known as the monkey tamarind. I recognize the name from a song by the psychedelic funk band Beginning of the End. The lyrics mention "a wild fruit that grows down Nassau way" but warn against eating it because it causes itchiness (presumably something you're supposed to pantomime as you dance to their song). I've always been curious about this shimmy-inducing forbidden fruit.

Jaitt hands it to me. Its exterior doesn't look too different from a jumbo-sized string bean. As I open it, the similarities end. The ice cream bean is filled with a snowy, sweet, cotton-candylike substance with hints of vanilla cream coursing through its translucent veins. It's like eating cloud. It's the most delicious thing I've ever tasted. I can see how it might make someone dance with joy.

"The thing about fruit tourism is that wherever you go it's always different," Jaitt says. "Nature is amazing all over the world. There are tens of thousands of different types of fruit that we know of so far. How many are in your supermarket? Maybe twenty-five different types?"

Pointing at his novelty-filled fruit bowl, he asks if I know where chocolate comes from. "Cocoa beans?" I offer, tepidly. "Okay, what's a cocoa bean?" he asks. I have to admit that I have no idea. He points at an orange football-shaped object. "That's a cacao. It's a fruit. All chocolate comes from the seeds of the cacao fruit. Wanna try?" He opens it up, and hands me some ice-cube-sized seeds surrounded by a white, gelatinous substance that tastes sort of like mangosteens. Just as Europeans used pepper as a currency, cacao seeds were used as money by the Aztecs. In the Middle Ages and Mesoamerica, money literally grew on trees. After I suck the flesh off, Jaitt explains how that seed is then roasted and processed to make chocolate. "People have no idea how their food grows and where it comes from," says Jaitt. "They just buy it in the supermarket."

As I prepare to leave, Jaitt hands me a couple of fruit magazines. Flipping through the pages of *Fruit Gardener,* seeing photos of fruit hunters posing with incredible fruits in obscure destinations, I realize that an entire unexplored community of fruit fanciers is being opened to me. I want to know more about them, to understand their passion, to travel with them, to keep tasting these fruits.

At the airport, I take out a handful of glassy purple jaboticabas, Jaitt's going-away present. Looking at them, I can't help thinking that fruits, so commonplace that we barely consider them, seem to be concealing some otherworldly mystery. For Joyce, the idea of the epiphany in the everyday was about "a sudden spiritual manifestation." Beauty, or truth or God, can exist within anything, particularly in the places that are so evident that we'd never think to look.

Foucault defined curiosity as "a readiness to find strange and singular what surrounds us; a certain relentlessness to break up our familiarities and to regard otherwise the same things." I took one photo of the jaboticabas in full focus; and then another out of focus, the globules dissolving into pure geometry.

These jaboticabas, full of promise, seemed to be pointing at revelations that I had yet to experience. Holding them, I felt like something miraculous had fallen into the palm of my hand, like the answer to a prayer I hadn't even realized I was making.

3

How Fruits Shaped Us

In the fruit trees are hidden certain of God's secrets
which only the blessed among men can perceive.

—St. Hildegard von Bingen

B ACK AT MY HOTEL, two wealthy, middle-aged ennui-wilted
Brits sitting next to me strike up a conversation. They're here on
holiday, but they haven't done much so far. "There are four-
hundred-year-old trees that are less bored than I am," says one of them,
puffing indolently on his pipe and affecting an extra-droopy countenance.

The dissipated gentlemen are sitting with their girlfriends, pneumatic
twentysomething twins dressed identically in yellow jumpsuits, hoop
earrings, pirate bandanas, and sideways ponytails. I ask them if they're in
the entertainment industry.

"Yes, they sing and they dance," explains droopy jowls.

"Our band is called Cherry Summer," exclaims one of the twins in a
Manchester accent denser than Marmite.

"Oh—why *Cherry* Summer?"

There's a pause. The other one rolls her eyes. "Umm . . . cuz we like
cherries?"

"Well," concedes Sir Droopalot, drawing deeply from his pipe, "that
pretty much says it all."

Having by now spent considerable time pondering the mysterious power of fruits, I wonder if it's that simple. Ever since Brazil, I've known that fruits make me happy, although I'm still trying to understand why. I enjoy being around fruits, especially when I can pick them off a tree and eat them. I try to start every day by eating fruits. When one of the twins asks why I'm so interested in fruits, I blurt out the first thing that comes to mind: "Because they represent everything that's wonderful in the world."

That's only partially true. Another reason we care about fruits is more selfish: without them, humans never would've happened. Tree-dwelling apelike protohumans first emerged somewhere between 5 and 9 million years ago. Fruits helped them evolve. Without the immense diversity of fruits, wrote Loren Eiseley in *The Immense Journey*, "man might still be a nocturnal insectivore gnawing a roach in the dark."

Fruits opened our eyes. Humans, alongside certain birds and primates, are part of a select group of species that can detect a difference between the colors green and red. Our 3-D photoreceptor eyesight stems from the need to notice red-ripe fruits in a sea of green leaves and foliage. Today, a red traffic light means "stop," which is exactly what we used to do in our primeval forests. The greens and reds are now part of the asphalt jungle, but their meaning hasn't changed much. In the same way, taste theoreticians speculate that humans first evolved a liking for sugary things as a way to distinguish between ripe and unripe fruits.

As our knuckle-dragging ancestors straightened into upright postures and banged out some crude tools, fruits were consistently part of prehistoric man's diet. Creeping out of the trees, we started eating berries and grains growing in the grasslands. Little by little, we made our prehensile way out of Africa and around the world. Until about thirteen thousand years ago, we were all hunter-gatherers devouring whatever we could get our opposable thumbs around—mainly acorns (the fruit of the oak tree). Humans are estimated to have eaten more acorns than wheat or any other fruit.

Figs, wheat, barley, peas and beans were the earliest crops domesticated at the beginning of the Neolithic revolution. By 4000 B.C., inhabitants of the Fertile Crescent were growing olives, dates, pomegranates and grapes. Land-owning Sumerians, Egyptians and Greeks enjoyed a slim selection of hard-to-grow fruits, which were widely cultivated during the Roman empire. Caesars returned from overseas victories bearing never-before-seen fruits as talismans. Planting throughout the empire, Romans brought seeds wherever they roamed. We tend to think of the apple as British, or even American, but it was spread by Rome, via the Caucasus.

At that point, many fruits were eaten dried or cooked because they were smaller, tougher and more sour than what we're familiar with today. Olives were brined and pressed to make oil. Grapes, occasionally eaten fresh, were primarily used for making wine. Figs, enjoyed straight off the tree in season, were also baked or preserved. Other fruits, if eaten, were usually processed before being consumed. Civilization was turning away from wilderness. The advent of food preparation, wrote Lévi-Strauss in *The Raw and the Cooked,* "marks the transition from nature to culture."

Only through human cultivation were fruits improved and selected for desirable characteristics: smaller seeds, increased flesh and refined eating quality. This discredits the assumption that wild fruits are tastiest—in truth, uncultivated varieties are often inedible. The wild peach is an acrid pea-sized pellet. Feral bananas are filled with tooth-shattering seeds. Untamed pineapples are full of gritty pebbles. Sweet oranges only arrived in the Mediterranean basin in the late 1400s. Corn is believed to have evolved from a minuscule grain called teosinte, slightly bigger than an earwig. It took thousands of years of human selection for teosinte to become the size of a human finger, then thousands more years to become the thick cobs we slather with butter today.

Not surprisingly, many fresh fruits were considered unhealthy by classical medical authorities. Pliny said that all pears are indigestible unless boiled or dried. Columella warned that peaches reek with "malevolent poison." Doctors counseled that apricots were to be eaten and then voided by vomiting repeatedly.

The physician Galen, whose second-century teachings prevailed for close to 1,500 years, cautioned against eating fruits, claiming they were troublesome in every way: they caused headaches, gullet distress, bad corruption, fevers and even premature deaths. The merit of raw fruit was as a digestive aid—more for purging than for enjoying, and best kept near the vomitorium. Galen believed fruits were, at their best, laxative. "We never need them for food, but only as a medication," he wrote. Galen used to make house calls and feed constipated patients pears and unripe pomegranates. He noted, with satisfaction, the remarkable fashion in which their bowels would soon evacuate. The Western belief that fruits served only a medicinal purpose lasted until the Renaissance.

The nomadic tribes that sacked Rome saw no need for agriculture, so they uprooted trees. Barrenness descended upon Europe. "It would not lift until forty medieval generations had suffered, wrought their pathetic destinies, and passed on," writes historian William Manchester.

In Asia, fruit culture burgeoned during the prosperous golden age of the Tang dynasty from the seventh to the tenth centuries, with the best fruits being grown in imperial gardens for the emperor and his intimates. During the Song dynasty, there was a widespread cult devoted to creating art about flowering plums. Fruits such as citrus, bananas, cherries, apricots and peaches all came from the Far East, by boat over the Indian Ocean, or by caravan over the Silk Road connecting China and Persia. Many others were endemic to the Middle East.

South of the Mediterranean Sea, fruits spread with the rise of Islam. As the caliphates expanded through North Africa, a number of new Asian fruits followed suit. Because Muslim doctrine forbade alcohol, vineyards were torn up and replaced with tree fruits. Europeans are indebted to Arabic civilization not only for nautical charts that helped them find their way around Africa and to the New World, but for a numerical system that gave birth to modern capitalism. Alongside calculus, they pioneered geology, astronomy and archaeology. They also taught Europeans to start enjoying fruits.

In the 1100s, the Crusades exposed Europeans to thriving fruit-growing regions abroad. Marco Polo's account of his travels in the Orient, filled with descriptions of magnificent pears, apricots and bananas, generated much excitement. "They have no fruit the same as ours," he wrote. At that time, it was thought that fruits and spices literally came from Paradise, believed to be located somewhere in the East. With the discovery of the Americas, conquistadores brought back mirabilia such as the pineapple, the papaya and the potato. Columbus made diary entries about his fruit finds: "There are trees of a thousand kinds all producing their own kind of fruit, and all wonderfully aromatic; I am the saddest man in the world at not recognizing them, because I am certain that they are all of value."

But among the majority of Europeans, Galen's legacy persisted, and raw fruits were avoided. They caused "crude and windie moisture." They aggravated melancholy. They affected the humors adversely. They were responsible for the tidal waves of deaths caused by infantile diar-rhea. In the fourteenth century, France's Eustache Deschamps blamed fruits for causing the plague and cautioned readers to avoid "fruits both old and fresh if you want to live a long life."

Unsurprisingly, colonialists stigmatized local crops wherever they set-tled, relegating once vital plants to oblivion. They also wiped out local populations, losing accumulated knowledge about indigenous flora.

The consumption of fresh fruits was minimal until the 1860s, reports Paul Freedman in *Food: The History of Taste,* noting that "raw fruits were regarded as dangerous in the Middle Ages and Renaissance, even if (or perhaps because) delectable." Sixteenth-century royalty were among the first Europeans to consider raw fruits as delicacies. Going against his doc-tor's cautions, Louis XIV boldly ate fraises de bois. Russian tsars used to send *estafettes* to gather wild strawberries in Lapland. Greengage plums were among the sunken treasures found in Henry VIII's flagship, the *Mary Rose,* submerged in 1545. Frederick the Winter king of Bohemia heated stoves year-round to produce fruiting orange trees in his castle at Heidelberg. King Charles II commissioned an artist to paint his portrait with a pineapple, the ultimate status symbol of his time. Father Athana-sius Kircher, in 1667, noted that pineapples have "such an excellent taste that the nobility of China and India prefer it to anything else."

Following the monarchs' lead, fruits became an aristocratic indul-gence. In 1698, Francis Misson de Valbourg wrote that, "Fruit is brought only to the Tables of the Great, and of the small Number even among them." Fruits were a demarcation between the upper classes and the lumpen proles. Haughty gentry used to carry around perfumed and spiced fruits called pomanders as a way of warding off nasty street odors. When assailed by the miasma, they'd stuff their noses into pomanders and escape into sweetness.

At that point, most fruits were still smaller and less juicy than they are today. Humankind had learned how to domesticate and propagate fruits, but we were only starting to understand how to breed them for desired traits. As fruit gardens were established, variety collecting became a high-society hobby. Orchards were emblems of wealth staffed by servants and full-time gardeners. They represented taste, refinement, even power. By selecting superior fruit, and growing these cultivars, eating quality improved. The Enlightenment's inquiries into the natural world yielded numerous treatises on fruit growing. Cabinets of curiosity, including

Ulisse Aldrovandi's celebrated *Wunderkammern* and Francesco Calzo-
lari's *Museum Calceolarium,* featured fruits amid myriad other naturalia.

By the late Renaissance, writes historian Ken Albala, "Italians were
clearly crazy about fruit in any form." Soon enough, European medics
started suggesting that fruits might actually be healthy. In 1776, doctors
described raw fruits as "the lightest most wholesome food we can eat."

The impoverished masses still ate whatever they could scrounge up. A
common peasant fruit was the medlar, a now-forgotten brown fruit that
must be dried (called bletting) before being eaten. It was nicknamed
"open-arse." Besides that, most people's dietary options were restricted
to gruels, porridges, turnips, cabbage, the rare salted meat and a little
bread. Until the early nineteenth century, with the exception of a few
landowners and noblemen, everybody around the world was dirt poor.
Life expectancy hovered around forty years.

People consumed fruits in beverage form. In the United States, the
majority of fruits were used for cider, perry or mobby (peach brandy).
Because drinking water was viewed as unsafe, fruit booze was what
everyone drank. Sir John Fontescue has pointed out that the English
were always drunk; they drank no water, except for religious purposes. As
U. P. Hendrick wrote, "fruit growing in America had its beginning and
for two hundred years had almost its whole sustenance in the demand for
stirring drink." Historians have long noted that it was a radical shift for
Americans to begin eating their fruit rather than drinking it. Only among
the affluent were high-quality fresh fruits appreciated. Washington, Jefferson and other landowning plutocrats respected fruits, discussed varieties in postprandial chitchat, and owned slaves that tended their
orchards. They were "gentleman farmers," men of independent means
who farmed for pleasure, as opposed to the vast majority of farmers who
produced food as a means of survival.

Until the industrial revolution, the North American population was
predominantly rural. People grew their own food. There was little fresh
fruit in the summer, and none in the winter. City dwellers had even
fewer fruits. Those that were sold took so long to get to the city that they
were often decomposing by the time they arrived.

As colonial conquests coughed up stimulants like coffee, tea and
chocolate, a hungry urban labor class emerged, its gaping maw clamoring
for calories. The price of sugar fell, making fruit-based preservatives,
jams and marmalades more widely available. In the nineteenth century,

writes Laura Mason in *Sugar-Plums and Sherbet,* fruit-flavored candy was "seen as an affordable substitute for fruit, at least by the poor." The convenience of cheap sugar-infused fruits facilitated our communal beguiling at the hands of artifice: we never had the real stuff to begin with.

Today most of the products on our supermarket candy racks are fruit imitations or derivatives: Swedish Berries, Jolly Ranchers and Skittles. Chocolate, as Jaitt pointed out, comes from the cacao fruit. Bubble gum used to come from chicle, the latex of the sapodilla tree, also known for its sweet chico fruits.

Aztecs had been chewing chicle long before the Europeans arrived. In the early twentieth century, American gum manufacturers hired thousands of South American bubble gum harvesters, called *chicleros,* to gather the sapodilla's sap. This industry died out after WWII with the introduction of petrochemical ingredients. Today, gum is made with a plastic oil derivative called PVA (polyvinyl acetate).

The advent of canning in 1809 furthered the availability of fruits. And even though the taste was a tad metallic, the innovation allowed fruits to be eaten year-round. In the mid-1800s, American authors like A. J. Downing chided farmers into growing edible fruits: "He who owns a rood of proper land in this country, and, in the face of all the pomonal riches of the day, only raises crabs and choke-pears, deserves to lose the respect of all sensible men." But as late as 1869, P. T. Quinn noted that fine fruits "are a luxury that can only be indulged in by the wealthier classes." It was around this time that better-tasting fruits started being cultivated on a wider scale in America. But there was still no way to ship these fruits to the expanding urban populations.

The shift from horse and buggy to locomotives facilitated fruit transportation, but farmers then found themselves needing to produce fruits that could survive long-distance shipping. The preeminent Georgia peach was the Elberta, sufficiently firm to make it to New York City without turning to gooey mush. Henry Ford's assembly line became the model for production. The advent of refrigeration, supermarkets and family automobiles abetted the urban availability of fruits—although taste quality suffered. An influx of 7 million Italians (mainly in between 1880 and 1921) also had a major impact on American eating habits and agriculture. Their love of produce was contagious.

Until the twentieth century, much fruit in Britain rotted on the trees. Adding to the fruitlessness was England's damp climate, unsuitable for drying fruits in the sun. In the 1890s, the apple became Britain's national fruit. The government started an "Eat more fruit" campaign. Although eating citrus was known to prevent scurvy, tens of thousands of soldiers succumbed to the disease as late as the 1910s.

The conclusive discovery of vitamins around the time of WWI marked a decisive shift toward considering raw fruits not only as beneficial, but as necessary. Still, fresh fruits disappeared during the twentieth century's world wars. Canadian families received rations of "raspberry" jam: it was, in fact, sweetened turnips dotted with wood chips to simulate the seeds. Matisse said that fruits were "more expensive than a beautiful woman" during wartime. Grapefruit only caught on in America after the Great Depression, because they could be exchanged for food stamps. Even then, people thought they needed to be boiled for hours before being eaten.

At the end of WWII, the British government allocated one banana to every child. Evelyn Waugh's three children were giddy with excitement on the great day their bananas arrived. As Auberon Waugh recalls in *Will This Do?* their joy was short-lived: the bananas "were put on my father's plate, and before the anguished eyes of his children, he poured on cream, which was almost unprocurable, and sugar, which was heavily rationed, and ate all three . . . He was permanently marked down in my estimation from that moment on, in a way which no amount of sexual transgression would have achieved."

Since the postwar era, fruits have enjoyed a spectacular boom, with new fruits continually appearing in supermarkets. Ted Hughes was twenty-five when he tasted his first fresh peach in London in 1955, the year he met Sylvia Plath. Kiwis arrived in the sixties. Mangoes and papayas turned up shortly thereafter. Fruits from South America, Asia, Africa, the Caribbean and the Middle East have been trickling into every one else's diet as waves of immigration and travel have exposed Westerners to exotic delicacies. Fresh figs arrived in Montreal a few years ago. Until 2006, only 5 percent of Americans had ever tasted a pomegranate, but that number is changing rapidly.

The past half century has seen a surge in availability, despite decreased quality. This was, perhaps, a necessary transition. In many ways, we're entering a golden age of fruits: never before have so many of us had access to such a wide range of fresh fruits, whether novelties or heirlooms. And the produce section will be getting even more interesting in

rs as innovative breeders and growers continue to focus on
s shoppers rediscover seasonality. We're only now starting to
the vast diversity that surrounds us. Forget about just red or
es: there are around ten thousand grape cultivars worldwide.
There are a hundred wild species of cherries, not including the thousands of cultivars selected over the years. There are 5,000 cataloged cultivars of pears. Over 1,200 cultivars of watermelon grow worldwide.
There are over six hundred different types of dates. As we'll see, there are
manifold reasons why we haven't as yet tasted these things. The point is:
the world of fruits is always evolving.

THE WORD "FRUIT" comes from the Latin *fruor,* which means "to
delight in," and *fructus,* which means enjoyment, pleasure and gratifica-
tion. When *fruc-* morphed into the Proto-Germanic *bruk,* it retained the
connotation of "to have enjoyment of." By the Middle Ages, however,
the meaning depreciated to "digestible," and by the sixteenth century it
signified "tolerable." Following the meandering stream of etymology,
bruk became the English verb "to brook," as in to endure. Although fruits
today are certainly built to last, they have also managed to retain their earlier connotation of bestowing bliss.

William Carlos Williams wrote about "a solace of ripe plums" consol-
ing a grieving old woman. For Cézanne, apples were a way of finding
inner peace. "Comfort me with apples for I am sick with love," pleaded
Solomon in the *Song of Songs.* Fruits were part of Einstein's simple formula for joy: "A table, a chair, a bowl of fruit and a violin; what else does a
man need to be happy?" Seventeenth-century courtier Nicholas de Bonnefons dealt with stress by hanging out with fruit trees. He believed that
fruits spread contentment: "One must confess that of all foods only fruits
win the prize for highest satisfaction." In the *Thousand and One Nights,*
bananas have a special appeal to mourning females: "Bananas . . . who
dilate young girls' eyes / Bananas! When you flow down our throats, you
don't collide into our organs ravished to feel you! / . . . And you alone,
amongst all the fruits, are endowed with a sympathizing heart, O con-
soler of widows and divorcées!"

There are theories suggesting a link between diet and teenaged angst.
Kids prescribed Ritalin for their attention deficit hyperactivity disorder
show significant improvement when they start eating fruits for breakfast.

Fructology is a system of fruit-based therapy that involves having your aura cleansed by a Life Fruit that corresponds to your astrological profile. The website www.thefruitpages.com tells of people who have conquered depression by eating fresh fruits on a regular basis. Andy Mariani, a San Jose orchardist who grows the best peaches in America, credits a nectarine given to him by his mother when he was dying from a debilitating autoimmune disease with giving him the will to survive.

The healing power of fruits is being substantiated by scientific research. Figs contain omega-3 and more polyphenols than wine or tea, citrus peel combats skin cancer, and kiwi fruits have blood-thinning properties similar to aspirin. Bananas relax us and alleviate depression thanks to tryptophan, a protein that increases serotonin levels. Cranberries are laden with phytochemicals that cure urinary infections and fight everything from kidney stones to cholesterol to ulcers. They also contain proanthocyanidins, called PACs, that surround harmful bacteria so they can't stick to our insides.

The supermarket fruits we eat contain negligible traces of protein, carbohydrates, cholesterol, sodium or fat (except avocados). Some have moderate amounts of dietary fiber. Where fruits excel is in their high levels of vitamins and minerals. Plants, and all fruits, contain a variety of chemicals, called phytonutrients, that are essential to human health.

The best way to absorb these nutrients is by eating a rainbow of fruits daily. Colors indicate different benefits. Anything pink—like watermelons, pink grapefruit or cara cara oranges—contains lycopene, an antioxidant that neutralizes harmful free radicals. Reds and purples—blueberries, cherries, apple skin, blood oranges and pomegranates—are indicators of anthocyanins, a flavanoid shown in a 2007 study to destroy cancerous cells without affecting healthy human cells. Orange fruits—papayas, mangoes and peaches—contain carotenoids that protect against heart disease and muscular degeneration. Yellow and green fruits—avocados, green grapes and peas—signal the presence of lutein and zeaxanthin, vital to human eyes.

Apple growers concocted the catchphrase "an apple a day keeps the doctor away" as a marketing scheme in the beginning of the twentieth century. Research today seems to support the idea: apples clean lungs, reduce asthma and cancer risks, and studies say they are better than toothbrushes at getting rid of the bacteria in our mouths. The act of peeling apples activates and improves our brain functioning, counters dementia and stimulates our creative faculties. A Yale University investigation concluded that the smell of spiced apples can prevent panic attacks.

Our ancestors understood that fruits, as part of the forest pharmacopeia, were full of medicinal attributes. Eastern herbal remedies include many fruits barely known in the West. China's melia fruits are used as painkillers. Sky fruits, from the Solomon Islands, help blood circulation and kidney function. A recent Bulgarian study of two hundred men suffering from impotence demonstrated that the dried fruits of *Tribulus terrestris* increase sperm production and motility.

Increasingly, the pharmaceutical industry is trying to understand fruits' medical properties. Black mulberries contain deoxynojirimycin, a chemical that combats HIV. Mars chocolates have launched a new line of medical products using cocoa to treat diabetes, strokes and vascular disease. Grapefruit, it's been revealed, can disrupt a variety of medications including antidepressants and high-blood pressure pills. Anise fruit's carminative properties earned it the name *tut-te see-hau*—meaning "it expels the wind"—among certain Native American tribes. It is also antiseptic, antispasmodic, soporific and a few seeds taken with water can cure hiccups.

We're all supposed to have a minimum of five fruit servings a day, ideally more. American per capita consumption limps around 1.4 servings per day. Potatoes and iceberg lettuce account for one third of all vegetables sold in North America because of their role in fast food.

Income directly affects the amount of fruits we eat. The more we have, the more we understand the complexities of health, and hence we turn to fresh produce. Affluence leads to diversified fruit consumption. On the other hand, fruits, while cheaper than cigarettes or alcohol, are too expensive for people on subsistence diets. Although bananas and oranges are certainly affordable, you can live without them. Very little fresh fruit has been available at restaurants, especially fast food restaurants, but now Apple Dippers are becoming the norm.

Fruits today play an integral role in preventative, wellness medicine. We seem to think that we're only now realizing how important fruits are, but in many cases we're only now rediscovering lost wisdom. As Hippocrates said, in one of the earliest of all recorded aphorisms, "Let food be your medicine and medicine be your food."

FRUITS HAVE INSPIRED countless innovations. Humankind's first ever work of art is believed to be a piece of carved silica in the form of an almond, or a seed, dating back two or three hundred thousand years to the lower Paleolithic. The Sumerians invented writing to document grain and fruit trades: the earliest cuneiform tablets were accounting records for agricultural administrators. Words like *logos* (meaning "word," "language" or "reason"), *legere* ("reading"), *lex* ("law") initially referred to things collected in forests, such as acorns. The first prose text in Latin was Cato's *De Re Rustica,* which talked about growing fruits near cities. Wheels came about as a way of transporting fruits with ox-drawn carts. Mulberries were the reason that humans invented paper, silk and forks. (Paper was initially derived from mulberry trees, without which silkworms couldn't spin their magic, and the berries were too messy to eat by hand—the way we ate everything else—so tines came about.) Our first bowls and containers were the fruits of the *Crescentia cujete,* otherwise known as bottle gourds or calabashes (which played a crucial role in the birth of agriculture in the Americas). The burrs on burdock fruits that get stuck in clothes provided the inspiration for Swiss engineer Georges de Mestral to create Velcro.

According to Greek mythology, the earliest musical instrument—the lute—was invented by Apollo, who carved it out of a melon slice. Initially, many musical instruments were made out of fruits. African gourds were fashioned into string and percussion instruments. Music stores still sell items like totumo fruit gourds, lacquered fruit maracas and seed-rattler anklets. Some of the first guitars and violins in the United States were slave-made stringed gourds. Humans have been singing about fruits since at least 1000 B.C., according to the Shi Jing, an ancient Chinese songbook that mentions seventeen varieties.

Fruits were pivotal in a number of scientific discoveries. We owe gravity to apples. Darwin's theory of natural selection was confirmed by his experiments with gooseberries. Robert H. Goddard was sitting in a cherry tree at age sixteen when he was struck with a vision of creating a device that could fly to Mars. He went on to become one of the fathers of modern rocketry.

We've used fruits in countless creams, cosmetics and cleansers. The illipe nut is used in lip balms. Avocado skin is used as a facial rub. Shea butter, used in ointments and lotions, comes from the fruit of the African karite tree. American beauty-berries, which are the violet color of Elizabeth Taylor's eyes, contain chemical compounds that can be used as insect repellent.

The lipstick tree of South America produces red nuggets called *achiote* that were formerly used as body paint. Today, these seeds are used as a dye called annatto that colors everything from butter to salad oils. Red carmine dye, or cochineal, comes from the pulverized corpses of small-scale insects that turn red after eating cactus fruits. Numerous other fruits also produce tannins used in paints, dyes and other coloring agents.

Fruits could replace many toxic cleaning products (most of which contain artificial scents like "fresh citrus"). Kaffir limes are used to wash hair in Bali. Jamaicans clean floors with orange halves. My Parisian friends do their laundry using soap nuts, the dried fruit of the Chinese soapberry tree (*Sapindus mukorrosi*). These berries contain saponin, a natural steroid that turns frothy and bubbly in water. I tried it: my laundry came out clean and smelling great.

Obscurely inspired, we've even used fruits as contraceptives. In medieval Europe, lemons were believed to neuter sperm the way citrus curdles milk. Casanova wrote of using hollowed out half-lemons as contraceptive diaphragms. Ancient Egyptians used orange halves in the same way, as do some misinformed modern teenagers. Unripe papayas can allegedly induce miscarriages and have been used in traditional societies as morning-after pills. In the sixteenth century, melons were prescribed to dampen sexual appetites. Herbalists from that period also claimed that medlars stay women's longings.

Our behavior toward certain fruits—so lusty, so full of desire, so fetishistic—stems from reproductive, survivalist instincts lodged deep within our subconscious. The fruits themselves encourage us by having evolved peculiarities that we respond to viscerally. And some people, I'm about to learn, have evolved their own peculiar ways of relating to fruits.

4

The Rare Fruit Council International

The sections of tangerine are gone, and I cannot tell you why they are so magical . . . there must be some one, though, who knows what I mean.

—M. F. K. Fisher, *Serve It Forth*

I N A SMALL Miami office crammed with books on the intricacies of fruit growing, the Fairchild Tropical Botanic Garden's senior curator of tropical fruit, Richard J. Campbell, is demonstrating how a chupa-chupa is opened in Peru. Using a pocket knife, he makes five vertical incisions at equal intervals around the oblong soft-ball-sized fruit, starting at the center of the nipplelike protuberance at the top. He then peels back the slices, so that the fruit's velvety brown-green skin blossoms open like a flower, revealing a startlingly bright orange interior. The glorious color contrast sends thrilling pulsations through my brain's pleasure center.

Campbell cuts five segments from the fiery orb, each of which contains a large seed. In his late thirties, Campbell has a crew cut and a frame that suggests daily workouts. He takes his kids shark fishing on weekends and has crisscrossed the globe looking for delicious new fruits that he's hoping to make available to a wider public. "With some of these ultra-tropicals," he says, handing me an orange chupa-chupa slice, "I think we can change the world."

Campbell demonstrates how to eat the fruit by sucking on the flesh. After all, it's called a chupa-chupa, which means "sucky-sucky" in Spanish. I place it in my mouth and suckle on it, releasing a flood of sweet juice flavored like mangoes, peaches, cantaloupes and wildberries. It's sensational.

Campbell knows how mind-blowing the fruit is. He also knows that it's a long way off from consumer acceptance. "You put that in a grocery store and who's going to know how to do what I just did?" he asks. It's his goal at the botanical garden to create a buzz that will eventually translate to commercial appeal. The chupa-chupa is one of a dozen or so ultraexotics that Campbell is working on bringing closer to produce sections. To that end, he has traveled down the Amazon River, from Peru through Colombia and into Brazil, trying to find the best chupa-chupas in the world. He has never, he says, been able to find any better than the one we're eating, whose seeds were initially collected at a market in Iquitos, Peru, in 1963 by his friend and mentor, William Whitman, who has grown it in his backyard ever since.

Whitman is the reason I've come to Miami. He's responsible for introducing hundreds of exotic fruits to Florida. This week is the opening of the William F. Whitman Tropical Fruit Pavilion, a thirty-eight-foot-tall crystal greenhouse filled with the finest known varieties of durians, mangosteens, dukus, bembangans, taraps and other rare tropical fruits. Campbell and Whitman spent several years traveling together to amass the pavilion's contents. Whitman has donated 5 million dollars to the Fairchild Botanic Garden to make the project a reality. Campbell, who is overseeing the donation, says that the knowledge gained through research in the pavilion will be a major step toward the commercial production of ultraexotics in America.

The pavilion's unveiling is going to be a gathering of America's most die-hard fruit enthusiasts. As soon as I heard about the launch, I called Kurt Ossenfort, and the two of us booked tickets to Miami. We're hoping to film William "The Banana King" Lessard, famous for growing the rarest of the rare bananas, and Maurice Kong, who has written numerous accounts of his globe-spanning adventures in search of nam-nams, candy-striped Malay apples, giant sapodillas, scarlet gac fruits, purple-fleshed guavas and Jamaican stinking toes. I'm also planning to say hello to fruit expert Bruce Livingstone, who owns www.tropfruit.com and changed his name to that of his favorite fruit: Santol. As a member of the

Thai Banana Club, he treks through forests to find near-extirpated banana varieties that he then brings home to conserve. Santol achieved fruit-world notoriety for discovering the Sarttra banana in northern Thailand.

The reason southern Florida is home to continental America's most devout fruit hunters is because its climate can accommodate the many rare subtropical fruits found abroad. The hobbyists gathered here travel to find new fruits that they can grow at home. When it was first settled, Florida had few fruits besides the coco plum and the seagrape. Decades of exploration have filled backyards with miracle fruits, jackfruits and egg fruits, used locally to make a milk shake called "egg-fruit nog." "The wonderful thing about tropical fruit is that there are amazing people involved," says Campbell. "There are all these guys around the world who keep in touch and travel together and have all these crazy stories."

WHEN OSSENFORT AND I meet William Whitman at his split-level Bal Harbour home, he is no longer the swashbuckling young man who rip curled his way into the Surfing Hall of Fame and introduced spear hunting to the Bahamas. Wearing a sailor's cap, the ninety-year-old zips around in a motorized scooter for seniors and gives us a tour of his garden, explaining where and how he first came across each of the different fruit trees.

He pats the trunk of a breadfruit tree. Whitman's love of fruits goes back to 1949 when he first tasted breadfruit in Tahiti. He recalls how, the morning after sleeping with a beautiful Polynesian girl, he opened his eyes to find his thatched hut filled with villagers silently gazing at him. Upon his return, after managing to grow a fruiting breadfruit tree in his own yard, he became a founding member of the Rare Fruit Council International, a Miami organization devoted to the study of unusual fruits. "I was primarily interested in uncovering new fruits that people had never seen or heard of and exploiting their potentials," he explains, reaching up from his wheelchair to pluck a bright yellow egg-sized charichuela. "The first time I found it, I thought, 'Gee, another curious-looking fruit.'" He hands it to me and tells me to take a bite. It tastes like lemonade-infused cotton candy. "And then I tasted it and went, 'Wow!' It was so delicious, I went cuckoo."

Independently wealthy (his father was a successful Chicago industri-

alist and his family owns Bal Harbour Shops, ranked "the #1 most pro-
ductive shopping center in the U.S."), Whitman has spent his adult life
embarking on countless tropical adventures and growing his fruit introductions on this estate in Bal Harbour.

We come to the largest miracle-berry shrub outside of Africa. I pocket
some of the small red fruits for later on, where a taste test confirms that
they're identical to the ones I tasted in Hawaii with Ken Love. Whitman,
the first person to grow a miracle-fruit tree in North America, tells us
that he no longer uses sugar, only miracle fruit. His wife, Angela, plucks
him one every morning before breakfast. Whitman explains that the
seeds were brought back from Cameroon in 1927 by the U.S. Department of Agriculture's fruit hunter, David Fairchild. He says this is one of
the few places on the mainland that I'll be able to taste it: after being perceived as a threat to the sugar industry, commercial production of the
fruit was banned by the Food and Drug Administration in the 1960s.

Nearby is a thorny cactus covered in dragon fruits. A little farther
down the path is a fruit that he describes as a pet project: the keppel.
Initially discovered at the deserted Water Palace of Indonesia, an erstwhile harem, the keppel was once used as an aphrodisiac by the sultans
and their odalisques. It was also rumored to make excrement and urine
smell like violets. Whitman decided to test this hypothesis: "I got paper
cups and every hour I'd do a little thing in a paper cup and smell it—
hell, it never smelled like perfume the way they say. To read their versions, they say that if you eat a couple of keppel apples, every time you
pee you can fill up a couple of perfume bottles—well, it ain't so." An
opportunity to confirm his findings unfortunately presented itself
when I went to wash my hands after our fruit tour and found an
unflushed stool in the toilet.

To facilitate his experiments, Whitman had his alkaline beach sandlot
dug up and replaced with six hundred truckloads of black, loamy, acidic
soil. He is galvanized by the challenge of finding and growing plants no
Floridian ever has before. One of the only Americans to ever successfully
grow mangosteens, he's also a member of the illustrious Explorer's Club.
His many fruit accomplishments are documented in a large coffee-table
book called *Five Decades with Tropical Fruit*. When he gives me a copy, he
signs it with a trembling hand, "Good luck to you, Adam, in your new
interest in tropical fruit."

"His single-minded obsession is truly unusual," says Richard Camp-bell. "He has a passion for fruit that no one else that I've ever worked with has. He wouldn't do anything else or dream about anything else, until he found that fruit he was looking for." Others describe him as "monomani-acal about tropical fruits." Even in recent years, when dementia started setting in, he still made trips down the Amazon in a wheelchair.

When he was younger, he'd bring his whole family on treks through jungle islands. "I can remember traveling with my dad and my mom, and we would travel with five surfboards and four unicycles and we would go exploring for exotic fruits," his son Chris recalls in an on-camera inter-view at the Whitmans' private museum atop Bal Harbour Shops. "A lot of times people would ask us if we were part of a circus or something." His kids loved growing up with a fruit-freak father. They assumed everybody had chupas-chupas in their backyards. Friends would clamor for a taste of the ultraexotics in Chris's lunch box, despite parents' and teachers' concerns the fruits might be poisonous.

Wrapping up the interview, we take the elevator down to the mall's ground floor, a retail palace filled with haute couture by the likes of Valentino, Chanel and Oscar de la Renta. Even in its early days, when U.S. shopping centers averaged one hundred dollars in revenue per square foot, Bal Harbour Shops brought in ten times that amount. William's brother Dudley suggests we head to lunch together. While he tries to interest Ossenfort in purchasing some film footage the Whitman brothers shot in Hawaii after World War II, Bill's sinewy sixtysomething wife, Angela, takes me aside and presses a twenty-dollar bill into my hand. I try to hand it back, but she insists. "You remind me of my son," she says, smoothing a hand over her coiffed platinum hair. "I know how hard it can be when you're starting out."

Angela chauffeurs us to the Whitmans' gated community in her white Cadillac. "We buy a new one every year or two," she says. I'm not sure if she's referring to the car, their house or gated communities in general. She mentions something about owning the streets we're driving on. I peer into her frosted sunglasses through the rearview mirror.

When we arrive at their country club, Ossenfort and I help Whitman out of the car. He struggles to stand up, his whole body shaking from the exertion. Over cheeseburgers, Chris explains how the family's favorite place to go fruit hunting is Borneo. "There's more tropical fruit per acre in Borneo than any other place in the world," asserts Whitman. He speaks fondly of the pitabu, which tastes like orange sherbet with hints of almond and raspberry. He's also a big fan of red-fleshed and red-shelled durians, and says the island is home to the world's finest mangosteens. As I explain how I'd like to travel around to document these fruits and the community of people who appreciate them, Whitman waves his quivering hand dismissively. "It can't be done."

IN 1898, a twenty-eight-year-old named David Fairchild oversaw the creation of America's first department of Foreign Seed and Plant Introduction. He spent his life on a series of fruit adventures spanning the world, and is responsible for bringing more than twenty thousand plants into the United States, including varieties of mangoes, cherries, dates and nectarines. He was an early champion of the mangosteen, predicting that it, and many other tropical fruits, would soon be available on suburban American tables. Campbell and Whitman, with their pavilion, are hoping to finally make that vision a reality.

Fairchild described himself as a "fruit bat." As batty as he was affable, his childhood preoccupation with nature kept him too busy to get into any mischief. He loved gazing at plant particles under microscopes, and once wrote that "a man could spend his life and not exhaust the forms or problems contained in one plate of manure." But it was a pair of pajamas that changed the course of his life.

Those fateful pajamas belonged to a fabulously rich globe-trotter named Barbour Lathrop. Giving his home address as San Francisco's Bohemian Club, Lathrop lived his days in exotic ports of call spending his vast family fortune. One morning in November 1983, aboard the steamer *Fulda,* the twenty-four-year-old Fairchild saw him decked out in pajamas and stared at him in openmouthed astonishment. (Pajamas, from Japan, were only then starting to replace nightshirts as the dominant form of sleepwear, and Fairchild, going abroad for the first time to study, had never seen any before.) Charmed by the young man's unquenchable inquisitiveness, Lathrop decided to fund his botanical trips to the Far East.

The two men ended up traveling the world together for the next four years. Fairchild called him Uncle Barbour. Lathrop called his protégé Fairy. After setting up the Foreign Seed and Plant Introduction for the

USDA, Fairy spent the next four decades hunting useful plants—especially fruits—around the world.

Whether having to drink water contaminated with dysentery or getting lost in impenetrable fever-infested forests, his misadventures were legion. Fairchild's junk caught fire in the South Seas. He "frequented most of the filthy places to be found in the world." When shipwrecked in Celebes, he came across one of fruitdom's great rarities: a hardened coco-pearl formed inside a coconut the way pearls form inside oysters. He ate dates in the souks of Fez and the oases of Algiers. The last descendent of the kings of Kandy taught him how to eat watermelon-sized honey jacks (far superior than regular jackfruit).

In 1905, he married Alexander Graham Bell's daughter Marian. Together they traveled, finding yellow raspberries in Padang and the square, angular fruits of the *Barringtonia speciosa* in Mozambique. In Siaoe, dozens of softly singing children followed the newlyweds around the entire island.

As he grew older, Fairchild started dispatching other fruit hunters to still unexplored regions. He sent Wilson Popenoe to Latin America, where he found billiard-ball-sized blackberries. (One of the largest berries known to man, a single Colombian blackberry comprises several mouthfuls.) Another of Fairchild's emissaries, Joseph J. Rock, was posted to the Orient to find the *kalaw,* a semimythical fruit thought to cure leprosy. Never before documented scientifically, Rock sought it throughout India, Thailand and over the mountains of Burma, fending off leopards, tigers and venomous serpents. In the Gulf of Mataban, he found one tree, but it was merely a cousin species to the *kalaw*. Finally, in a decrepit encampment near the upper Chindwin, he stumbled upon wild *kalaws* just as a rampaging elephant and a typhoon simultaneously ravaged the village, stomping tents and washing away trees.

For his Asian envoy, Fairchild was looking for a peripatetic type "who could tolerate all sorts of physical discomforts and walk thousands of miles where no roads existed." Frank Meyer's Herculean physique and reputation as a long-distance hiker got him an interview, but the twenty-nine-year-old was so nervous that his profuse perspiration caused the colors on his striped shirt to run together. Still, Fairchild took a shine to the sweaty man, and from 1905 to 1918, Meyer chased fruits through dust storms and across frozen mountains. He finessed donkeys over precarious bamboo bridges spanning chasms; he was attacked by murderous brigands; and he trekked over Siberian glaciers so cold that milk froze in his cup before he could drink it.

He visited many places that had never seen a white man, let alone such a large, strapping beefcake. He was often asked to ripple his muscles and crowds would gather to watch him bathe. In the pear-growing district of Tongchangdi, villagers scrambled onto rooftops to catch a glimpse of him. In other areas, natives were so afraid of the hulking foreign demon that he could only placate them by sitting down and eating fruit to show that he was just like them.

In photographs, Meyer is bearded and tempestuous, with a gnarled walking stick and puffy leather knee pads. But beneath his sheepskin coat, he wore pinstriped three-piece suits. He probably slept with his monocle on. "Work is to me what medicine is for sick people," he wrote. "I withdraw from humanity and try to find relaxation in plants."

He brought back seedless persimmons and melting quinces from Tianjin, red blackberries from Korea, famed pound peaches from Feicheng, paradise apples from Elizavetpol, kiwis from Ichang. In Feng-tai, he stumbled upon his ticket to immortality: Meyer lemons. While his lemons are more available than ever before, Meyer himself disappeared from the deck of a steamer crossing between Wuhan and Nanjing on the night of June 1, 1918.

David Fairchild spent his final years in Miami, and his legacy lives on at the botanical garden named in his honor. For today's fruit hunters, who congregate at this luxuriant plant sanctuary, Fairchild's books are sacred texts. "His books put a wanderlust in me that was almost uncontrollable," explains Campbell. William Whitman says he found Fairchild's books so interesting that he decided to do the same thing with his life: discover fruits and bring them home. In his autobiography, Whitman claims that he has introduced more fruits to the U.S. mainland than any experimenter since Fairchild.

ON THE MORNING of Whitman's ribbon-cutting opening, a buzzing crowd between the ages of forty-five and ninety-five has assembled in front of the greenhouse. Tables piled high with artfully arranged tropical fruits are off to a side. Smiling people are drinking coffee, wondering aloud whether we'll be getting any liquid sunshine, and discussing rare fruits. I overhear a passionate debate over the merits of the black sapote, a fruit that one attendee describes as tasting like chocolate pudding. His detractors compare it to axel grease, cow patties and wet turds.

Another topic is an upcoming group trip to Asia to be led by Chris Rollins, the director of Homestead's Fruit and Spice Park. I ask Richard Wilson, owner of Excalibur Nurseries, what motivates his fruit excursions. "Knowing that there are fruit out there that nobody has brought home and if you can, you'll be another Bill Whitman." He's such a beloved figure that every June 7 is Bill Whitman Day in Miami.

In his floppy white sun hat, Murray Corman of the Garden of Earthly Delights Nursery looks like an older Gilligan from *Gilligan's Island*. Corman explains that Whitman has exposed him to "unotherworldly" flavors. The strangest thing he's eaten, Corman says, is the *Bromelia penguin,* a wild orange pineapple relative used as a meat tenderizer, something that only dawned on him when his taste buds started dissolving. He describes the sun sapote as having a wonderful flavor reminiscent of cooked sweet potato and the texture of a scrub brush. "It's quite delightful," he adds. "Literally you have to rip the fruit from the fibers to enjoy it."

I ask if anyone can taste these fruits domestically. He makes an offhand remark about smuggling them in. Smuggling? Yes, he says, certain enthusiasts are known to occasionally smuggle home rare cultivars. When I press him for details, his goofy Gilligan grin fades, and he says, curtly, "I don't think that would be an appropriate conversation to have."

Although most of Miami's fruit hunters follow importing protocols, some skirt regulations. For that reason, government officials at the United States Department of Agriculture have started conducting armed raids on rare fruit growers, bursting into their backyards with attack dogs. Richard Wilson tells me that he has pressed charges against the USDA for a violation of civil rights when a squadron of machine-gun wielding agriculture agents raided his nursery and accused him of smuggling in seeds. "They came in here like the goddamn Gestapo," he says angrily. "It was like they were gonna save us from terrorism. Six agents burst in and started rifling through everything trying to find illegal seeds. They orchestrated it like it was a big drug raid. It scared the shit outta my wife, not to mention my customers. They photographed stuff, and confiscated seeds. They thought I had smuggled 'noxious weeds' in from Asia. They were after an illegal seed that was one-sixteenth of an inch long. My palm seeds were an inch long. They didn't know a noxious weed from a palm tree seed. They wouldn't know a noxious weed if it grew in their butt. I got a five-million-dollar business here; you think I'm gonna grow an illegal plant and screw it up? They took three of my palms—palms not known in cultivation—and they killed them. They all died. I talked to the head authority in the U.S. on CITES [the Convention on International Trade in Endangered Species of Wild Flora and Fauna] and he didn't know his ass from a hole in the ground. I was falsely accused. I never even fathomed anything like this illegal plant-and-seed smuggling interdiction service existed."

It's becoming clear that some of these fruit lovers can be pretty intense about the objects of their desire. The Fairchild garden's director Mike Maunder, who seemed a bit freaked out by the crowd, describes the fixation with tropical fruit as a prime example of horticultural fanaticism. "The fruit people are driven by that real sense of exploration," he explains. "They are going into the Amazon, going into New Guinea, and looking for unusual fruit." Maunder points at a couple of gentlemen sitting on the grass and suggests that Ossenfort and I chat with them about how they "love fruits, live fruits and collect fruits from all over the world."

One of them, a bearded fellow in a leisure suit and a green Rare Fruit Council baseball cap, introduces himself as Har Mahdeem. He is a specialist in *Annonas,* a genus of trees and shrubs that produces large custard-filled fruits like atemoyas, guanábanas, cherimoyas, llamas and bullock's hearts. "For rare fruiters, finding a new fruit is a great adventure," he says, with a wide smile. He's been on several collecting trips to Guatemala and Central America, looking for orange, pink and red *Annona* fruits to bring back for his employer, Zill High Performance Plants in Boynton Beach, Florida.

Mahdeem was born in Michigan but grew up in the Amazon basin near the city of Manaus, where his missionary parents started an agricultural school. "I soon acquired a nickname: *Bouritirana,* which is the name of a small, worthless fruit," he says in a singsong voice with a indeterminate, vaguely Southern twang. He was given the name because he'd asked someone if he could eat it. "That was my perpetual question as we walked through the forest: 'Can this be eaten? Can this be eaten?'"

In recent years, he legally changed his name to Har Mahdeem, which means "Hills of Mars" or "Martian Mountains" in Hebrew. I ask him why he likes fruits. "They're eye-catching, nose catching, taste-bud catching," he says, laughing. "There are thousands of plant species that produce edible fruits. No one has ever tasted even a fifth of them."

Sitting beside Mahdeem is an elfin man in his sixties named Crafton Clift. Scrunching his face in deep concentration, he says he has no idea why he loves fruits so much. Perhaps it has something to do with wanting to be a child again. He also has a nickname: Graftin' Crafton. They call him that, pipes in his teenaged nephew Scott, sitting nearby, because "he grafts *way* too much."

Grafting is a means of propagating a plant by cutting a branch from one tree and sticking it onto the trunk of another tree. For the most part, this technique is used to clone a desired variety in order to create more of the same fruits on other trees. Clift's passion—or compulsion—is grafting together as many different species as possible and seeing what comes out. "I like making fruits—creating them," says Clift. "It's like the excitement of breeding a Great Dane for the first time."

A graft from a certain species can take hold only on a related species. Apple branches can't be grafted onto orange trees. But you can graft different citrus varieties together so that the same tree bears lemons, limes, grapefruit, tangerines, kumquats and citranges. A Chilean farmer recently made international headlines with his "Tree of Life," covered in grafts of plums, peaches, cherries, apricots, almonds and nectarines.

Sometimes, a graft mutates into a different fruit. This is what's known as a sport. Perhaps one in every million or two grafts will spontaneously mutate and develop different characteristics than its progenitor. If the bud mutation is interesting, yielding a different colors, or flavor, the sport can then be propagated itself. Exposing branches to mild doses of radiation is also done in order to expedite mutations.

When humans first started grafting around 6000 B.C., it was seen as a form of magic. "Up to the end of the middle ages, grafting was considered a secret by the initiated and a miracle by the public," wrote Frederic Janson. Some believed that, for a graft to hold, it was necessary for a man and woman to make love in the moonlight. At the moment of climax, the woman was to secure the graft between the tree and its new limb. The royal fruits of the Tang dynasty were only grafted by the wizardlike hunchbacked gardener of Ch'Ang-An. English philosopher John Case ludicrously claimed that grafting a pear tree onto a cabbage was "a wonderful fact of art."

In the early days of American life, shady types took advantage of the public ignorance of grafting. There were mountebank grafters who stuck branches onto trees with wax—by the time you realized you'd been had, months had passed and the vagabond gardener was long gone.

Poets argued about the ethics of grafting. "He grafts upon the wild the tame; / That th' uncertain and adulterate fruit / Might put the palate in dispute," wrote Marvell. "And in the cherry he does nature vex, / To procreate without a sex." Abraham Cowley saw in grafting a hint of Godlike power: "Who would not joy to see his conqu'ring Hand / O'er all the Vegetable World command?" Religious leaders frowned upon grafting as tampering with divinity. It was, according to the Talmud, an abomination.

Shakespeare, in his comedy *The Winter's Tale,* argued that grafting is indeed natural: "Yet Nature is made better by no mean / But Nature makes that mean. So, over that art / Which you say adds to Nature, is an art / That Nature makes . . . This is an art / Which does mend Nature— change it rather, but / The art itself is nature." (One wonders whether he'd apply this same argument to transgenic modification or cloned meat.)

The distrust eventually subsided. Grafting became a passion for historical figures like St. Jerome, the Ostrogoths' King Theodoric the Great and George Washington. As with any other aspect of fruit growing, it attracted intense devotees. Feng-Li, a Chinese diplomat in the fifth century B.C., abandoned his career when he became consumed by grafting. The North American Fruit Explorers handbook even warns that grafting can become "a bit overwhelming . . . an intense activity bordering on obsession."

Clift doesn't seem to have heeded the handbook's warning. "As long as he can graft, he's happy," says Richard Campbell. Because Clift has been caught in the Fairchild Tropical Botanic Garden climbing trees and grafting different species together, security now forces him to leave his knives and grafting implements at the door. "He *has* to propagate," explains Campbell. "I know how it is: you find something wonderful and you want to graft it. But I do it in my yard—not in the botanical collections of the United States of America. If someone showed him a strange fruit, he would just drop everything and follow it and lose his job."

Indeed, Clift seems to float in and out of projects. Periodic dispatches in newsletters mention his "retirement" from one garden or how a "dream job of looking for new fruits" didn't pan out because of management conflicts. "People like Crafton aren't accepted by the establishment because they're different—they're odd," says Campbell, explaining that Clift had been banished from the Peace Corps. At one point, he says, incredulously, Clift was offered a job working with fruits in Costa Rica. He decided to drive down from Florida. In Guatemala, his suitcases were stolen. In El Salvador, he was sleeping by the road when someone robbed the clothes off his back. He continued to drive in the nude. Abandoning the car in Nicaragua, he proceeded to walk the rest of the way to Costa Rica, living on jungle fruits and trekking naked through forests for weeks.

"He walks through the world in a complete naive wonderland of fruit," says Campbell, recounting how, at another point, Clift had been hired by a wealthy Thai family to create the world's biggest tropical fruit garden. He was forbidden from sending seeds to other fruit enthusiasts, but was caught in the act. His hands were to be chopped off as punishment. Luckily, he managed to escape, fleeing Thailand.

On the lawn of the Fairchild Botanic Garden, Clift says that his main problem with the Nong Nooch gig was more existential. "I was asked to collect all the tropical fruits of the world," Clift explains. "The director said, 'How come no one ever thought of this?' I didn't even answer him. I thought: 'You have no idea how many tropical fruits there are in the world. And even if you collected them all, you still would only be at the starting place, because then you start selecting to find the best ones, and grafting and hybridizing.'"

Clift becomes visibly overwhelmed when speaking about the limitlessness of fruits in nature. "It's like you open a room and you see all the tropical fruits there and you think 'That's it?' No. Whether it's the jungles of Borneo or the Amazon, every door that you open opens into a larger room. And this is before we start hybridizing and recombining the genes or even selecting to get the biggest, the juiciest. It's day one with tropical fruits. A tremendous variety are just now being named and discovered. As long as we have any forests left at all, we will find new fruits. Give us a few centuries to cultivate these things and they're going to be very different. In my lifetime things have changed so much."

I SPEND THE rest of the morning asking guests at the pavilion's opening about where to find their favorite fruits. Their answers are as limitless as Clift's jungle rooms. The best white strawberries are Purén, Chile's *frutillas blancas,* although other sweet specimens grow near Istanbul and Royan. The finest dates in the world grow in the *khlas* groves of al-Mutairfi in Saudi Arabia. The white loquat is only available in the last weeks of May in Suzhou, China. The cannibal tomato is used in Polynesian headhunter sauce. The butter fruit is a creamy fruit best enjoyed in the Philippines, as is the yellow, lycheelike alupag and the kalmon. According to Florence Strange, when opened, the kalmon "reveals a beautifully spiraled gelatinous center with tentacles radiating out from the top." The honied mohobo-hobo is mixed into an orange porridge called *mutundavaira* in Southeastern Africa. The Milky Way yamabôshi is an oblate scarlet fruit noted for its papaya flavor in Japan. The *Pandanus edulis* is a red, brown and yellow fruit from Madagascar that resembles a cluster of wide-bottomed bananas growing fused together like a pineapple. The best pomegranates are found in the Iranian towns of Kashan, Saveh and Yazd. In Lebanon, unripe plums, grapes, apples, lemons and green almonds are eaten with salt as snacks. "It's like your senses are being invaded," is how one of them longingly describes it.

Richard Campbell says that his favorite fruit is any of the world's fruits at their peak in their right environment. "When you sit in an ancient lychee orchard on Hainan Island, China, with an old man with the wrinkles of life in his face and he hands you a lychee that his family has grown for a thousand years and you eat that fruit—that's your favorite fruit."

Hundreds and thousands of cultivars spiral off from the tens of thousands of umbrella species that bear edible fruits. As Graftin' Crafton points out, nobody knows precisely how many fruits exist, because new ones are still being discovered and others go extinct before ever being documented. Still others materialize through artificial or natural selection.

Fruits grow in every climate and in every region. Different elevations, microclimates, and land conditions all contribute bursts of diversity. The skin-searing heat in desert areas produce fruits, as do the murky marinations of bogs, swamps and marshlands. Cloud forests, where plants are continually enshrouded in fogs and clouds, abound with species that can't grow anywhere else. The wildest diversity of fruit life is to be found in the resplendent cummerbund on the tuxedo of planet Earth in between the tropics of Cancer and Capricorn. Temperate habitats, which house hundreds of species, cannot compare to the richness of tropical rain forests, where tens of thousands of species coexist.

A visitor to the Sub-Saharan region will encounter numerous fruits, such as the plumlike caura, a number of berries including the sône and the tekeli, and the hazelnut-sized cobaï fruit, said to be so delicious that, when in season, no other food is touched.

Of the over 750 species of fig, some grow in desert regions and others ripen underground. The Violette de Bordeaux fig is filled with a molten flesh that tastes like raspberry jam. The Panachée or tiger-striped fig is yellow with green stripes. Its red interior has a flavor likened to strawberry milk shakes. Fig expert Richard E. Watts has written in *Fruit Gardener* how "a few private collectors now have most of the rare figs in Southern California." One of them, Jon Verdick of www.figs4fun.com, grows more than three hundred different types. Where can people taste quality figs? "The simple answer is to buy them from me (grin)," writes Verdick. Or grow your own. Figs once grew in the hanging gardens of Babylon. Before the tenth century, the best fig was the Sbai, which still grows in Israel. According to certain apocryphal scriptures, figs in Eden used to be the size of watermelons.

From Manchuria to Manitoba, prairie habitats also play host to numerous fruits. Blueberries were first domesticated in New Jersey's Pine Barrens, where they now grow in massive quantities. Even severe arctic climates produce fruits. Doyenne du Comice pears, with their distinctive Chanel No. 5 fragrance, do well in gardens north of Toronto, Canada. Orchards of apples, plums and peaches thrive in Kazan, Russia, despite spending months buried under snow. Certain species of kiwi grow in Siberia.

Alaska is the one state that doesn't ship produce to the Hunts Point Terminal Market in New York. When I inquired why they don't carry Alaskan fruit, the market's executive director gave the sort of exaggerated pause normally directed at imbeciles or little children. "Because nothing grows there," she snapped, frostily.

In fact, Alaska is home to numerous vegetables, not to mention apples, blueberries, raspberries and other fruits. After I listed cloudberries, nagoonberries, salmonberries, mouse nuts, beach asparagus, wild cucumber, kelp, dulse, rhubarb, spiked saxifrage, silverberries and serviceberries—which are mixed with sea purslane and reindeer fat to make Eskimo ice cream—the Hunts Point executive director gave another pause, and then said haughtily that she hadn't heard of any of them.

Thousands of tantalizing fruits that never make it to North America or

Europe are eaten everyday across the globe. Even if we knew they existed, importing fruits over borders is a process fraught with botanical, economic and geopolitical challenges. Complicating matters, most little-known fruits don't produce abundant enough crops to merit shipping. They're also only in season for a brief period. And with many fruits not cultivated on a mass scale, quality varies widely from tree to tree. All of which is exciting to fruit hunters, but anathema to supermarket supply chains.

Fruits have trouble traveling between different temperature zones. Just as temperate berries deteriorate when shipped to the tropics, tropical fruits taste less flavorful in colder regions. Fruits are highly perishable, and weeks of transport only aggravate matters. Commercial fruits aren't picked ripe; they're picked whenever boats are scheduled to depart. These unripe fruits often turn mushy and even start to ferment in transit.

For all these reasons, the research undertaken at the Whitman pavilion could prove instrumental in ushering in a new world of northern fruits. As the greenhouse's fledgling trees grow tall over the coming decades, the fruits produced will be studied in order to be grown and sold on a wider scale.

"Come along, Eric," says Angela, confusing me with her stepson. She leads me to the Jean duPont Shehan Visitor Center, explaining how members of the duPont family—the famous chemical manufacturers—are in attendance. At the postinauguration lunch reception, baskets of red Hawaiian rambutans are passed around the banquet hall. They've been allowed into the mainland because of a new irradiation technology. The buzz around the room is that X-rayed fruits will soon be allowed into America not only from Hawaii, but from Southeast Asia, South America and other tropical destinations. (The first shipments of legally imported, irradiated, Southeast Asian mangosteens were finally allowed in three years later, in 2007). There's a consensus that irradiated fruits are never of the highest quality. To truly know these fruits will still require traveling to their places of origin, as the fruit hunters have always known.

Part 2

Adventure

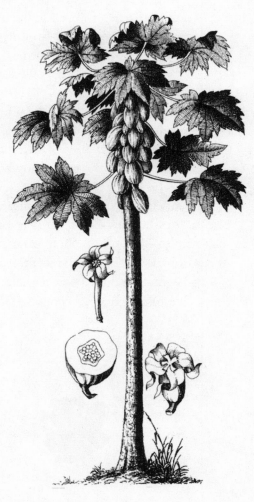

Carica papaya, West Indian pawpaw

5

Into Borneo

Note that at the farthest reaches of the world often occur
new marvels and wonders, as though Nature plays with
greater freedom secretly at the edges of the world than
she does openly and nearer us in the midst of it.

—Ranulf Higden

O NE MONDAY JANUARY MORNING in 2007, New Yorkers
awoke to a nauseating odor of sulfur and rotten eggs. It was as
though a gigantic fart had rippled through Lower Manhattan.
The lingering gassiness, described in the media as "awful," "nasty" and
"ominous," caused the evacuation of schools and office towers; 911 circuits were jammed, service on commuter trains and subways was suspended, and a dozen people were hospitalized. "It may *just* be an
unpleasant smell," said Mayor Bloomberg, seeking to calm citizens. Fingers were pointed at New Jersey's chemical factories; others speculated
that it was a leak of mercaptan, the substance that gives natural gas a
noticeable funk. In the end, the source of the stench was never identified.

I think I know what might have caused it: durians.

The durian is the most odoriferous fruit in the world. Containing
forty-three different sulfur compounds, including the same ones found
in onions, garlic and skunks, this spiky fruit befouls any enclosed space.
Its penetrating smell is intended to attract animals like orangutans, tigers

and elephants. Fine in a rain forest, but not in downtown Manhattan, where a Dumpster full of gaseous, decaying durians could wreak havoc.

Think I'm exaggerating? A few months after our trip to Miami, Kurt Ossenfort and I held a durian-tasting party at his apartment in New York. Opening them unleashed such an invigorating gust that it felt like being in one of those old cassette commercials where the sound of the tape blows your hair back as you hold on to your seat. The two durians we'd picked up in Chinatown were so powerfully fetid that, as we feted the fruits, the rest of the building was evacuated due to a suspected gas leak. Unbeknownst to us, durian vapors were moseying down the hall and through the elevator shaft to other floors. Concerned tenants grabbed their valuables and took cabs uptown or waited anxiously in the deli around the corner. We found out only when the worried superintendent and an official from ConEd showed up to pinpoint the source of the leak.

We weren't the only Manhattanites sharing durians that year. The *London Review of Books* mentioned a similar durian party attended by Susan Sontag, Lou Reed and curators of the Museum of Modern Art. Upon learning that durian would be served, Fischerspooner's singer reportedly "emitted girlish little squeals."

Other reactions aren't as enthusiastic. It has been compared to rotting fish, stale vomit, unwashed socks, old jockstraps, low-tide seaweed, a charnel house, sewage in a heat wave, pig shit and baby diapers, turpentine, a disinterred corpse clutching a wheel of blue cheese and French custard passed through a sewer pipe. Eating them is said to be like eating your favorite ice cream while sitting on the toilet. The ones we ate in Manhattan tasted like undercooked peanut butter–mint omelets in body-odor sauce. Smoldering burps resurrected the flavor well into the following morning.

Not surprisingly, breeders are developing scentless varieties. Even so, true durian aficionados love its pungent aroma. In Malaysia, there's a saying: "When the durians come down, the sarongs come off." Durian-scented condoms are successful in Indonesia. Just as Southeast Asians are repulsed by raw-milk Roqueforts, Westerners have trouble understanding the attraction to durians. Many cultures enjoy foods on the edge of rotten, whether it's Sardinia's *casu marzu* cheese filled with wriggling maggots, Iceland's *kæstur hákarl*—decomposing shark meat—or dessert wine made from grapes gnarled with botrytis mold.

Durian flesh, isolated from the scent, is actually quite sweet. An oft-repeated durian adage is that it smells like hell, but tastes like heaven. As fan site durianpalace.com puts it: "Imagine the best, most delicious, and sensuous banana pudding, add just a touch of butterscotch, vanilla, peach, pineapple, strawberry, and almond flavors, and a surprising twist of—garlic??!!" One of the first positive descriptions in the English language noted wafts of "incongruities" such as cream cheese, onions and brown sherry.

It is banned in many hotels and public spaces throughout Asia. In Singapore, signs on subway stations warn that durian carriers face a five hundred Singapore dollars fine. Durianpalace.com thinks banning nature's grandest pudding is "a hopeless rule just like outlawing farts, when we know that it is such a pleasure and everybody's secretly doing it." Aviation alerts have been caused by passengers transporting the fruit in their luggage. Citing "grossness," not safety, Virgin Airlines manager Brett Godfrey canceled a 2003 flight in Australia because of a castaway durian. "It just is the most pungent, disgusting smell," he said, suggesting it belongs in an outdoor dunny.

It's also advisable to abstain from alcohol while eating durians. Pairing them entails serious bloating. Jerry Hopkins, in *Extreme Cuisine,* mentions a newswire report of a "fat German tourist who devoured a ripe durian, followed by a bottle of Thai Mekong rice whisky, then took a hot bath and exploded."

I also felt like exploding after our tasting party—not surprising given the quality of the fruits. If they're malodorous when fresh, consider how bad it gets when a low-grade variety dethaws on a congested Chinatown sidewalk after spending months with freezer burn. More than 10 million dollars are spent on frozen durian in the United States annually. They are all substandard, repugnant even, compared to a freshly picked Thai Golden Pillow. In the nineteenth century, as he was readying his own theory of evolution by natural selection, Alfred Russel Wallace claimed that durians were "worth a voyage to the East to experience." Heeding Whitman's assertion that getting a proper ultraexotic experience still requires travel to the tropics, I booked a flight to the pulsating heart of the durian kingdom.

ONE OF THE largest islands in the world, Borneo is a geopolitical triumvirate comprising Malaysia's Sarawak and Sabah, Indonesia's Kalimantan and the tiny sultanate of Brunei. Countless fruits that are unavailable elsewhere thrive in this remote center of endemism. In recent decades, the island's rain forests, formidable hotbeds of diversity, have suffered devastating losses from logging. Fortunately, many of the region's most valuable fruits are being grown and studied by conservationists.

Voon Boon Hoe is a botanist who has long been William Whitman's and Richard Campbell's contact in the region. Clean-cut and thin, with a salt-and-pepper mustache, Voon has spent his life studying durians at Sarawak's Agricultural Research Center.

As we walk through the research orchard outside the capital city of Kuching, the island's variation becomes increasingly apparent. There are bloodred, chartreuse, yellow and orange rambutans. Even sweeter are their relatives, pulasans, whether scarlet or green. We eat cluster bombs of dukus and langsats, tangy-sweet detonations of citric perfection. All of these fruits would be phenomenally popular if they ever became available in the West. Gungungs taste like wild strawberries accentuated with raspberry coulis. Dabais are like oversized purple olives that melt in the mouth, releasing an oily nectar. The star-shaped *Baccaurea reticulata* has a vermilion exterior and a milky interior with a glossy white seed the size of an avocado's. Just gazing upon it puts me in a trance.

The island boasts more than 6,000 indigenous plant species. Borneo's genetic pool is so rich because it's a relatively undisturbed ecosystem that wasn't affected by the crystallized extinctions of the Pleistocene Ice Ages. It gets up to seven meters of annual rainfall. Like other equatorial rain forests in South America and Africa, Borneo is bursting with extraordinary life-forms: tiny owls, deer the size of mice, flying lizards, Oz-like flying apes, luminescent mushrooms and colorful fungi resembling coral reefs. Butterflies suckle on human sweat while expelling creamy secretions. Voon points out a little bird called the black-breasted fruit hunter.

Clouds of mosquitoes trail me wherever I go. Although I'm taking anti-malarials, the travel clinic warned me that there's no vaccination against insect-borne dengue fever. Each time I'm bitten, I imagine some rare tropical virus coursing through my bloodstream. At least there's no sign of the other pests I've been warned about: tiger leeches that gorge on blood, sharp-toothed ants that swarm up legs and man-eating crocodiles.

While hiking through durian groves, Voon and I hear something plummet off a tree. Voon scampers through the tall grass like an excited little boy. Proudly bearing a spiky fruit, he explains that in Malaysia, the best durians are freshly fallen (as opposed to Thailand, where they're cut off trees before falling). We open it right up. It's a *kuning* durian, says Voon, handing me a piece of salmon-colored meat. The interiors of *kunings,* unlike the better-known yellow-fleshed varieties, range from neon orange to deep carmine. This one has an intoxicatingly nutty, almondlike taste, and only the faintest odor. It's infinitely better than the putrescent stink bombs found in Manhattan.

There are twenty-seven species of *Durio,* most of which are native to Borneo, including centipede durians, mini durians with almost imperceptible seeds and even a naturally occurring odorless variety called *sawo.* While most durians have spiky green husks and yellow flesh, some varieties, like the *Durio dulcis,* have bold red exteriors. One *kuning* tree in Dalat is so big that it takes fifteen people with outstretched hands to encircle it. The seed is said to have been given to the owner's grandfather by a ghost in a dream.

As we walk on, we come to a tarap tree. Voon sends up an assistant to cut off a large, soccer-ball-sized fruit that has already been half-eaten by a flying fox. "That's how you know that it's ripe," Voon explains. Brushing away some insects, he hands me a chunk. It tastes like a fully constructed dessert. The juicy white cubes of flesh fuse a custard's richness with a cakelike powderiness. The whole thing seems topped with vanilla-spruce frosting. The sweetness is heightened by a jungle creature's stamp of approval. There's a sense of togetherness, something ineffably primal, about sharing a fruit with a winged canid.

By now, I'm so transfixed by all the fruits that I've stopped visualizing myself hospitalized from mosquito-induced fever. We come to a tall mangosteen tree covered with fruits. In Borneo, Voon explains, people climb trees to eat mangosteens straight off the branch the way North Americans do with apples. He suggests I try it out. Climbing up, I perch on a branch laden with leathery, brown, apple-sized fruits. I throw a couple down to Voon, who demonstrates how to open them using his thumbnail to pierce the thick leathery skin. Twisting off the top, I discover the interior filled with perfect snow-white sections.

At the end of the day, Voon gives me a perfumed chempedak to take home. Filled with honey-sweet orange chunklets, it's an army-green fruit the size of a rugby ball. After taking a few bites, I leave it on the nightstand of my third-floor hotel room while I go out for the evening. Returning after dinner at a nearby hawker stall, its scent overwhelms me the moment I enter the lobby. Sneaking it outside via the rear entrance, I sit down in a nearby vacant lot and pull the fruit's sections apart. As the moon wavers in the thick haze, I dig in. Its flavor has improved since the afternoon—it seems to be at its apex of ripeness. The taste is somehow familiar, yet elusive. With every bite I try to place the flavor. Then it hits me: Froot Loops!

Tasting it triggers a recollection of how, as a child, I'd use my allowance to buy boxes of Froot Loops. I'd sneak away and covertly eat bowlfuls in bed under the covers while reading *Archie* by flashlight. Soon all that remains is the skeleton of a chempedak at my feet. I've eaten it all, my hands tearing it apart, the fleshy globules offering themselves to me. I can still feel fructose crystals coating my teeth like icing.

THE FIRST EVER official fruit hunt was organized by the Egyptian Queen Hatshepsut, who sent ships to the land of Punt in East Africa to retrieve seeds and plants in the fifteenth century B.C. Whenever galleons blew into shore, local tribes offered fruits as welcoming presents. Columbus was served custard apples upon landing in the Americas. Natives presented Cortez with unnamed oddities. Captain Cook was given breadfruit "the size and shape of a child's head . . . reticulated not much unlike a truffle." Structuralist anthropologists like Claude Lévi-Strauss were greeted at Amazonian deltas by flotillas of small boats containing exotic fruits. Even in relatively recent times, cargo cults in Papua New Guinea have been reported to brandish bananas at pilots.

After Columbus was welcomed with pineapples in Guadeloupe in 1493, the pineapple became a symbol of hospitality on gateposts and house turrets in Europe. They were adopting the aboriginal tradition of placing pineapples at entrances to welcome visitors and signify friendship. Five hundred years later, real estate agent William Pitt, specializing in Connecticut mansions, uses it as the company's logo, as does high-end cookware store Williams-Sonoma. Certain Super 8 Motel owners advertise themselves as Pineapple Kind of People because pineapples are "the pinnacle of perfection in the hospitality industry."

Fruit hunting has a storied legacy. One of the earliest European flora explorers was William Dampier, a distinguished pirate who swashbuckled through the Isthmus of Darien and down the Colombian coast. After being stranded in a canoe hundreds of miles from Sumatra, he somehow made his way back to England and managed to reinvent himself as a traveling botanist. His 1697 book *A New Voyage Round the World* was filled with fantastical fruits, which led to his being sent on several more official plant-hunting escapades.

John Tradescant gained notoriety with his 1621 expedition to procure the Algiers apricot. After joining a privateer's fleet sent to capture Barbary pirates off the coast of Algeria, he returned with numerous hitherto undocumented varieties of stone fruits. On subsequent voyages, he brought home white apricots and "an exceedingly great cherye" now known as Tradescant's black heart. In England, Tradescant discovered a new variety of strawberry, the Plymouth, growing on a rubbish heap in South Devon. It is noted for having miniature leaves that grow where the pips (or achenes) are usually found. All of his introductions were grown in "the Ark," a magnificent garden near Lambeth. Even though he had anosmia (no sense of smell), he and his son were renowned for the quality of their melons. They are also believed to have overseen some of the earliest cross-pollinations, although they never set down their findings in a verifiable manner.

Dry science didn't turn these adventurers' cranks. Following Tradescant, plant hunting became a craze among young men intoxicated with the idea of exploring new worlds. They were happiest dashing off into the unknown—their diaries are filled with accounts of momentous discoveries made after enduring punishing hardships: getting lost, eating raw vulture flesh, trekking through rain forests shoeless and clothesless, being sucked into whirlpools for hours on end, fending off rabid buffalo attacks, falling with their horses into hippopotamus wallows, having all their hair eaten by rats while sleeping and negotiating with armed, xenophobic natives ready to stone foreigners for trespassing on their sacred farmland. In 1834, David Douglas, who discovered the Oregon grape, died in a pit intended to trap boars in Hawaii: he fell in and was promptly gored by a wild bull. David Fairchild's protégé Wilson Popenoe's own wife died from toxic shock after eating an underripe ackee.

Fruit discovery was, for a time, a sure-fire status booster. Fruits were named after successful voyagers like Sweden's C. P. Thunberg, who managed to penetrate Japan's borders to pinch the barberry—now known as *Berberis thunbergii*—and bring it back to Europe. The feijoa, or strawberry guava, is named after Spanish explorer and botanist Don de Silva Feijo, who found them in Brazil. The kumquat's Latin name is *Fortunella,* after plant hunter Robert Fortune.

The French intelligence officer Amédée-François Frézier was spying on the Spaniards for the French government when he came across Chilean strawberries in 1714. Frézier—whose name, coincidentally, derives from the French word for strawberry—realized that the berries were more valuable than any state secret, and took every precaution to bring them home. The voyage took six months, and he almost died sharing his dwindling rations of fresh water with the plants, only five of which survived. These *Fragaria chiloensis* fruits supplied DNA that, when crossed with the Virginian strawberry, created the modern strawberries we eat today.

To introduce a seed into one's homeland was a way of enlarging its range and widening the horizons of human knowledge. As Thomas Jefferson said, "The greatest service which can be rendered any country is to add a useful plant to its culture." This rationale was how Jefferson justified smuggling rice out of Italy (risking execution) and hemp seeds out of China.

POPENOE DESCRIBED the fruit hunter's methods bluntly: "You go to some forgotten town and hire a native boy. Then you buy three animals—horses, mules or maybe camels—one for yourself, one for the boy and one for the baggage. Then you head for the back country and keep going until you reach a place where there are so few cooking vessels that you can pay for a night's lodging with an empty tomato can. Then you go to the village marketplace and watch like a hawk for everything brought in to be sold. You almost make love to the natives. You get invited to dinner. And finally you get the plant or seeds of the plant you want."

Even in the early twentieth century, it wasn't about plunging into the untracked wilds. It was about getting to the market. In general, the best varieties of fruits were found long ago and have been selected and bred for desirable traits over generations. Ethnobotanical work still necessitates firsthand surveys of forests, mountains, plains or valleys. Institutional fruit trackers have access to helicopters and parachutes, and use GPS and radar devices to hone in on their targets. For the most part modern fruit hunters don't bother much with the virgin forest; they hire local guides to bring them to private farms, orchards, agricultural departments, botanical gardens, nurseries, herbaria, laboratories and those rural markets.

As Richard Campbell puts it: "You fly into the closest airport, drive through town, peer into backyards, and find the guys who know where the best stuff is. Then you say, 'Hi, I'm from America, I'm crazy and I want to look at your mameys.'"

All fruit people are propelled by the notion of protecting biodiversity, whether through documenting wild species or propagating the plants at home. Bringing rare fruits home is a safety net against the pitfalls of extinction. Susan and Alan Carle have spent nearly three decades undergoing extensive collecting expeditions into endangered forests in order to protect disappearing species that they grow on their Australian property, called the Botanical Ark.

Harold Olmo, known as the "Indiana Jones of viticulture" for his grape-collecting adventures in Afghanistan and Iran, once spent three days having his car pulled out of a twenty-five-foot gorge by nomads using camel-hair ropes. The grapes he collected and bred played a major role in creating and sustaining California's wine industry. Afghan botanists recently obtained cuttings of fruits he'd collected that ended up disappearing from their native soil.

Roy Danforth and Paul Noren are Christian missionaries who have set up a tropical fruit preserve in the Ubangi region of the Democratic Republic of the Congo. Their Loko Agroforestry Project is devoted to preservation and reforestation, primarily with fruit trees intended to provide sustenance. I considered visiting them, until I read reports from otherwise staid botanists explaining that traveling to this source to interview them would entail hiring a helicopter gunship and a personal militia.

As Popenoe said, the best place to start looking for fruits is the marketplace. Anything worth growing or eating invariably ends up at these central congregations. Alongside uniting all of a region's bounty, produce markets are also social affairs where visitors meet locals, where friends touch base and where strangers bond over shared interests. I can spend hours at a market in a state of pure contentment, transfixed by the abundance and overstimulated by the contact with fellow humans and nature.

With all the raw ingredients waiting to be transformed into culinary splendors, markets are imbued with a sense of promise. They're harbingers of delights to come, colorful way stations on a journey to further happiness. Looking at the vast quantities on display, we feel a sense that we'll never be hungry again. We also love markets because the food there is real. It's fresher, of a higher quality, than food at the grocery store. This is especially true of North American and European farmer's markets.

There's something different about tropical jungle markets, full of native harvests dragged in from remote tributaries, perhaps never to be seen again. Something else attracts us to those markets, something darker. It may be the unruly nature of the negotiating, the shadowy codes of commerce, so different from the orderliness of supermarkets. There's also a sense of danger, of chaos, at tropical markets. Animals are slaughtered in front of your eyes. The smell of death hangs in the air. Such markets are more like intermediary zones between civilization and wilderness.

In the past, Western markets were just as strange. Medieval marketplaces in Europe and New France were more than shopping stops: they were places where justice was meted out, where bailiffs made legal announcements, where bards and fiddlers entertained travelers, commoners and the privileged. They were execution sites. Joan of Arc was burned alive in the marketplace (although her heart, it's said, remained impervious to the flames). The rabble jostled each other to catch a glimpse of the burning cadavers, to inhale the smoke of corpses.

At Kuching's Sunday market in Borneo, you can find everything from blowpipes and feathered arrows to meter-long green bean pods to hairy plant parts out of nightmares. As soon as you enter, balmy effluvia rising off the sidewalk clamp down on your sinus cavity. Duku husks, rotting coconuts, innards, transparent sacks containing muddy liquids and the pointy notes of overripe durians combine in an ague-inciting bouquet. Fermenting vegetables and dough cooking in rancid oil give off an additional sour aroma that actually gets caught in the back of the throat. The smell molecules are so powerful you can taste them. Carnivorous pitcher plants mottled with red veining and unpleasant prickly hairs are placed in plastic pouches where their open mouths gasp for breath. Writhing, roiling masses of fat sago worms are sold out of tree trunks. Men shout death threats at one another over corridors littered with squid tentacles. Dozens of bananas, from thumb-sized to over two feet long, nestle among sweet soursops, mountains of chilis, bright pink guavas, sea legumes and brown, dusky, powder-covered obelisks.

I'm the only white person here. Some vendors smile at me; most, however, treat me indifferently, shouting and barking orders to buy their products. I'm a terrible bargainer, but the pushy vendors seem to enjoy my comic attempts at negotiating (perhaps because they invariably come out on top). Everything is so tempting and magical—not to mention fantastically cheap. I leave carrying way too many bags of fruits and vegetables, many of which I have no idea how to eat or prepare. I arrange them into neat piles on the desk in my hotel room, appreciating their symmetry and colors, and occasionally taking cautious tastes of something possibly poisonous. I gorge on dukus, rambutans, soursops, mangosteens and durians, knowing that I may never again taste them in their native habitats.

Over the next week, I devour a guidebook Voon's given me called *Indigenous Fruits of Sarawak,* making a checklist of all the fruits that I still want to taste. There's the thumb-sized *keranji papan,* with its sweet orange caramel-flavored pulp. There are several varieties of *tampois* with pearly, transluscent interiors that make me almost tremble with desire. And then there's that *pitabu,* William Whitman's beloved blend of orange sherbet, almonds and raspberries. As I make my way through the rural food centers, I learn that many of the fruits I'm craving aren't actually in season now.

Voon's wife tells me that, one time when they were hanging out with Graftin' Crafton Clift, she said, "You guys are always talking fruit, fruit, fruit. Isn't there anything else?"

"*Is* there anything else?" deadpanned Clift.

In fruits, there's always something newer, better or rarer. It's the pursuit of infinity. Naively thinking I might somehow be able to sample everything, I start planning a return visit. Speaking to growers, they explain that different fruits ripen every month. There is no single period where a visitor can taste all of the fruits; the only way to taste them all is to spend an entire year in Borneo. Even then, there'd be many I wouldn't be able to taste because they're so remote. As I lie in bed, feverishly contemplating these fruits, the phone rings. It's my girlfriend, Liane. As I go on about all the different fruits I've been tasting, she says that it sounds like my subjects aren't the only ones lost in a naive wonderland of fruits.

To alleviate my gnawing desire, I eat dozens of durians in hawker stalls, marketplaces and restaurants. Whenever I catch a whiff of durian in the night air, I head over to the nearest vendor to eat a chunk, sold on a Styrofoam tray, which makes it look like chicken breasts.

I find myself succumbing to the same pensive, dreamy, glazed-over look I see come over other durian eaters. At first, I thought I was simply hypnotized by the flavor, a kind of custard perfection. Gradually, I'm starting to suspect that the act of eating it sets in motion certain ancient mechanisms in the brain.

At Voon's house, this precognitive process again makes itself felt when I taste a jackfruit. This time, there's none of the fear I felt with Ken Love in Hawaii. Licking jackfruit goo off my fingers, I feel like I've triggered some pathways in hitherto cauterized memory banks. Not only does tasting fruits bring us back to childhood, it brings us back to earlier evolutionary moments. Enjoying these fruits instills a sense of kinship with the ancestral humans who needed them to survive in the forest. Seeing, tasting and encountering durians, taraps and jackfruits seems to reanimate a primitive dimension of our subcortex, making our pulse quicken the way it would've if we were swinging through the trees eons ago.

According to Nancy J. Turner, an ethnobiologist specialized in aboriginal ecosystems and plant resources, "foraging for wild crops satisfies some instinctive yearning left over from man's evolutionary past when this occupation was essential for survival." When I first saw that Brazilian paradise nut tree, covered in muffins, it jolted me with a hardwired sense of excitement. The same neural circuits flash to life when I taste these jungle fruits: it's not only a sense of hope, it's an intimation of self-preservation, the knowledge that we'll remain alive for another day. That hunting and finding these fruits can activate deep centers in our unconscious minds is another example of biophilia, the love of life.

At the same time that I'm experiencing this bewildering neurochemical reaction, I'm constantly being presented with dismaying evidence of overlogging. The deforestation of Borneo has reduced the tree cover by more than half over the past fifty years. Driving through the country's heartland, I see decimated, defiled, embarrassed landscapes. The hillsides are covered in bald spots. Deep swaths are cut through the wilderness. Vast expanses of rain forest have been razed for palm-oil plantations. Nearly 13 percent of Malaysia is now covered in squat palm trees. Their orange, peach-sized fruits contain palm oil, a vital commodity used for cooking and biofuels. With the cost of food having increased by 37 per-cent in 2007 (and palm oil by more than 70 percent), families have started hoarding oil. Riots are breaking out in protest against shortages, so more and more of the forest is being converted to palm oil production. Fume-spewing timber trucks make up a hefty percentage of the traffic. The scars of human interference hang about the mist-enshrouded mountain-tops like shafts of light in a darkened, smoky room.

Deforestation, besides eliminating the planet's best defense against global warming—carbon-munching plants—also releases vast amounts of heat-trapping carbon dioxide into the atmosphere. Losing these forests is suicidal. Sixty acres of rain forests vanish every minute. That's 32 million acres a year. One percent of African forests disappear annually. "The Amazon loses an area the size of New Jersey every year to clear-cutting and timbering," reports the *The New York Times*.

Here in Borneo, it's muggy and overcast every day. An ominous brown cloud lingers over the entire island. Voon tells me that this haze sticks around for months at a time, the result of forest fires in Indonesia.

The devastation of these ancient forests has tragic consequences. Nomadic tribes, like the Penan, who used to survive on wild tree fruits can no longer maintain their traditional lifestyles. As their feeding spots have vanished, Malaysian officials have focused on assimilating them. Abdul Raham Yakub, a former chief minister of Sarawak, has said, "I would rather see them eating McDonald's hamburgers than the unmen-tionables they eat in the jungle."

While I'm in Kuching, the front-page story in *Malaysian Today* tells of a twenty-one-year-old former tribesman who was shot while picking wild durians in the jungle. After he crawled through underbrush for four hours, his grandfather found him and rushed him to the hospital. The bullets were surgically removed, and he ended up in stable condition, but the shooter was never found.

My hotel, the Telang Usan, is owned and managed by members of the Orang Ulu tribes. The front desk clerk tells me she can take me to a grove she knows that grows the best langsats. As we're driving out to the coun-tryside, she explains that she is a descendent of the Iban, also known as Sea Dayaks, a once-ferocious tribe that practiced head-hunting. At her village, a little painted sign bears the motto: "We believe in infinity." In a longhouse, I drink some *tuak* (a local rice liquor) with her family, who are now pepper farmers. Her grandmother stands up on a grid and demonstrates how to separate the green pepper berries from their stems by crunching them beneath her feet.

On my last day in Borneo, I fly to the erstwhile domain of the White Rajahs, Sarikei, for a Farmer's Day celebration. Outside the plane's window, plantations of oil palms stretch to the horizon. These fruit trees, with their bounty of edible oil, are the stars of Farmer's Day, which ends up being way less traditional than I had imagined. Voon promised that there would be many indigenous seasonal fruits to sample at this gathering, and indeed there are. Concomitantly, the event feels so corporate. Biotechnology companies hand out pamphlets explaining how their products will help maximize agricultural returns. The event's logo is a chemistry beaker sprouting a little green leaf. The theme is better living through agroscience.

Looking at some photos of the Penan, I see them eating other fruits that aren't even in the guidebook Voon gave me. It's saddening to think that, as the forest recedes, riches are being lost forever; yet, as I've learned on this trip, massive abundance still surrounds us. As Borneo's wilderness vanishes, there also exists an array of cultivated fruits so broad that it's impossible to taste them all.

At the end of the day, I join Voon and his colleagues at a folk dancing performance in an open-air stadium. The traditional forest dances seem out of place on this amplified, metallic stage. The crowd, wearing button-up shirts and dress shoes, watches the dancing and then stands up to sing along to an anthem. It's a catchy song with a syncopated Casio beat and couplets about embracing progress and cyber technologies: "Thank you for the way the old things were done," they sing. "Now it's time for the modern age."

6

The Fruitarians

Naught but fruit doth ever pass my lips . . . Eat of this fruit; believe me, it is the only true food for man.

—H. Rider Haggard, *She*

BANGKOK MEANS "village of wild plums." The city's main food market, which reaches full stride in the middle of the night and is over by dawn, is more like a megalopolis of wild plums.

Rising at 4 A.M., I flag down a market-bound *tuk tuk* driver. As soon as I embark on his motorized Thai rickshaw, he turns us around in a full circle, and then puts the pedal to the metal only to slam on the brakes a meter later, inches away from a parked truck. He then nearly runs over a bald pedestrian with a bushy black mustache. As we race over speed bumps, I fly around on the backseat wondering what to hold on to, the only choice being metal railings on the *tuk tuk*'s exterior that'll crush my fingers in the seemingly unavoidable event of a collision. Instead of letting my digits cushion any impact, I just grip the seat as my body flops around like a spawning salmon. We speed right up to enormous trucks and then pass them with millimeters to spare. Eschewing traditional lane changes and signals, my driver prefers to climb up a car's tailpipe and then swerve around it like a motorized bat.

I can smell the market before we arrive. At first it's pleasant: scents of basil, lemongrass, gingers, turmerics, galangals, mounds of freshly ground curry powder, heaps of coconut shavings. Once inside, however, the smells become so overwhelming that they tickle the back of my eyeballs. This is not the fragrance of guavas. This is the raw, cutting odor of the jungle, the gash of the tropics, the fetor of equatorial darkness, the essence of everything Western civilization glosses over, dyes and tries to not think about. The epicenter of this olfactory swamp seems to be a corner where crates of eviscerated frogs are piled next to barrels containing thousands of squirming crabs. The sharp odor of the splayed amphibians, their steaming organs perfuming the night air, mingles with the decomposing crabby emanations to create a stench nobody should ever have to experience again.

A smiling man chopping chilis gives me a moon fruit—a flat, yellow persimmon-like fruit that shoppers smell to block out the market odors. Wandering around with my nose buried in the moon fruit, I try to stay out of the way of porters, their loads toppling over on broken market alleys. As I jot down some observations, a car hits me—not fast enough to break any bones, but enough to shake me up. Feeling light-headed, I leave with a backpack full of mangoes, salaks, santols, rambutans, jambus and mangosteens.

By now, the sun is about to rise and the market is slowing down. A motorcycle taxi offers me a lift home. We drive so fast the helmet levitates inches above my head, barely tethered by the cord around my chin. Traffic lights don't seem to matter or make any sense. Neon signs blur into LCD cuneiforms. Closing my eyes as we race through the darkness, weaving in and out of traffic at 120 miles an hour, I imagine my body splayed across the road like a dissected market frog.

I'M ONLY PASSING through Bangkok, on my way down to a couple of islands in southern Thailand. Still curious about the fruitarian Punatics I heard about in Hawaii, I've arranged a meeting with the legendary durinarian Shunyam Nirav, creator of durianpalace.com. He spends half the year in a two dollars per night beach hut on a remote stretch of Koh Phangan. Getting there first requires making it to the island of Koh Samui. From Bangkok, I book an all-night train ride through central Siam.

Still itching from a sleepless night spent battling mosquitoes in my humidor of a train car, I arrive at my hotel on Silver Beach. Waiting for me is a note from Nirav suggesting that, while in Koh Samui, I meet up with his friend Scott "Kiawe" Martin. Not quite sure what to expect, I hitch a motorcycle ride to his home that evening.

Kiawe and his Thai girlfriend, Ta, live in a dilapidated beach bungalow surrounded by majestic palms and gramophone-like flowers. When I pull up, Kiawe, a tanned, rugged and handsome American, is lying in a hammock on the small front porch. He spends the rest of the evening in the hammock, lying back or bolting upright depending on his enthusiasm for the topic at hand. I sit opposite him on a wooden bench. Ta, wearing a headband and a tank top, is rather shy and spends most of her time indoors.

Soon after I arrive, a lanky twentysomething fruitarian named Jameson pulls up on a scooter. Taking off his helmet, he says that he heard I was interested in durinarians. "Pretty much all I eat is durians," Jameson says, shaking out his long blond hair. "Have you heard of the durian trail? There's a group of people who basically travel around following ripe durians so they can always find them in season."

We eat dinner—a crateful of wild local durians—as Jameson tells us about the first time he saw durians at a market. "They were cracked open, and I could smell them," he says, his hair twitching. "I picked one up. I didn't know if I'd die from eating it, but I tasted it—and loved it. So I backed my truck up, filled it with sixty durians and ate them for the next three days."

"How many durians do you eat at a meal?" I ask, feeling sated after eating two.

"Ten at a sitting," he says, and proceeds to demonstrate.

Kiawe's and Ta's diet is even stranger. Although they join us in eating some durians, they explain that their main course is yet to come. Alongside fruits, they also eat raw meats, many of which they dry themselves. Kiawe calls it "the caveman diet."

"We eat original food," explains Kiawe, "the way primal man ate before the advent of fire." The proper term for this diet, he says, is instincto anopsology.

Regular sushi or beef tartar pales in comparison to the raw meats they prefer. Kiawe describes one instincto who dries rabbits in the trunk of his car in order to eat the "beneficial maggots." He also explains his belief that getting sick is actually a body's way of getting healthy. "All illnesses are just cleansing reactions," he says. "It's your body's way of expelling toxins."

When Ta brings out a platter of chicken bones, fish ligaments, meat jerkies and other air-dried meats, I'm pretty certain that I don't want to taste any of it. They tear into the food with almost exaggerated pleasure. Kiawe holds a dried chicken out toward me.

"Will it give me a 'cleansing reaction'?" I ask, politely declining his offer. The prospect of getting ill—sorry, *cleansed*—on an isolated Thai island isn't high on my list of priorities.

"It might give you a cleansing reaction," answers Kiawe, unhelpfully.

"He's nervous," says Ta.

"He should be," says Kiawe.

"You should be," agrees Ta.

I pick up a dark chicken carcass. There's barely any flesh on it because of the shrinking caused by weeks of air-drying. I put a brittle tendon into my mouth. It has the texture and flavor of masking tape.

"What do you think?" asks Ta, testily.

"I really like plain things," I say, diplomatically.

"Yes!" agrees Kiawe, lurching in the hammock. "The bacteria is the seasoning. Aged meat is full of enzymes—like cheese or wine. Beef in the fridge for three months is great. You know how cheese gets mold? Beef gets it too, and it tastes just like cheese. We look like a bunch of cavemen, but it's really sophisticated stuff."

The rest of the dinner consists of Jameson and Kiawe explaining their vision of creating a fantasy durian orchard with thousands of wild and heirloom varieties. All they need is five hundred thousand dollars for their "Save the Wild Durian Project." I tell them that it sounds like a wonderful project. Kiawe places his hand on my shoulder and tells me that I have a special power: my curiosity.

After clearing away the bowls of bones, Ta asks me to follow her into the woods nearby. She brings me to a bush full of ripe specimens of a purple blackberry-type fruit covered in what appears to be crystallized fructose. It looks exactly like the gummy Jujube candies I used to eat as a child. Perhaps the candy's designer copied it, or maybe its shape has somehow been lodged in our collective unconscious from ancient times. I pluck one, and peer at its glistening sugar-frosted surface. The moist, meaty sections open up to reveal a blood-red raspberry jam interior. It tastes gritty, almost sandy, with a molasses flavor. Ta holds up a mirror. My tongue and teeth have turned completely black. "That's our favorite fruit," she laughs, assuring me that the effects will wear off by morning time. "We don't know what it's called, but we love it."

My blackened tongue wags good-bye to the original foodists. Jameson gives me a lift back to my hotel on his scooter. On the way, we pass a natural geyser that has erupted out of the sand. We pull over to look at the fountain. It is ringed by a lunar aureole, a sort of gray moon bow in the gloom. Jameson tells me I look like a zombie with my black mouth. Nearby, some fruit bats are munching on bananas. "They come in with six-meter wingspans over the water looking like pterodactyls," says Jameson.

After he drops me off at Silver Beach, I lie down on the sand under a large tree. It seems to be pulsating with faint lights. Am I hallucinating? Suddenly, one of the glimmering lights lands on my chest. It's a firefly. The tree is full of them.

THE FOLLOWING MORNING, a wizened old woman with two rows of lower front teeth hobbles into the lobby. Stopping at my table, she opens a suitcase containing strange poultices, wood chips, translucent orange liquids, brown pellets and what appears to be a dirty tropical truffle snaked-through with green and red sprouts. I point at it, and the traveling witchdoctor shakes her head and starts laughing. It's not for me, apparently.

Paying my hotel bill, I set off for the ferry on the back of another motorcycle. The driver points out sites of interest. We stop by a temple called Wat Khunaram, which boasts a macabre tourist attraction: a dead monk sitting upright in a glass case. Although he died in 1973, his body hasn't ever decomposed. He wears an orange robe and sunglasses to cover his empty eye sockets. His final request was that his body be put on display as an inspiration to future generations to follow Buddhism and escape the chains of suffering.

After docking on the island of Koh Phangan, I pass through Tong Sila, a saloon town with sandy dirt roads that's like a Thai replica of Dodge City. Soon enough, I arrive at Shunyam Nirav's beachfront hut, the one with a bough of purple bougainvillea hanging over the balcony.

A skinny, sarong-clad fifty-five-year-old with curly blond-and-gray hair comes out to greet me. As we sit on the stoop, Nirav opens a chempedak, careful not to stain his green sarong. The fruit is delicious, again reminding me of Froot Loops.

Nirav explains how he became hooked on durians when traveling to Bangkok in 1989. "I liked them immediately," he recalls. "Wow!" My eyes opened up, and I was totally enthusiastic right from the start. My girlfriend at the time was like 'Get that thing out of here.'" In the intervening years, Nirav has written songs and poems about the fruit, as well as the following haiku: "Ecstatic flavor/Spiky fruit's luscious pudding/Nature's grandest food."

Nirav has been on durian-tasting trips all over Southeast Asia and is part of what he calls "a real connoisseur scene." When he speaks of certain top-pedigreed Malay durians he's tasted in far-flung villages, words fail him. "They were just like . . ." he says, his eyes closing, his eyelids fluttering, and his hands rising in the air and waving slowly in total submission. And then, he snaps back to reality. "Mentally, I keep a one to ten scale. Some of these were thirteens."

For Nirav, eating durians is a spiritual practice. His name, he says, means "Emptiness Silence" in Sanskrit. The name was chosen for him by his guru, Bhagwan Shree Rajneesh, otherwise known as Osho (or the "Rolls-Royce Guru," due to his vast collection). Born Robert James Palmer, Shunyam Nirav became his legal name in 1990.

As a young man, Nirav had been a member of est, Werner Erhard's personal transformation seminars, later known as the Forum, or Landmark Education. On June 7, 1973, at the age of twenty-one, he attained enlightenment. "But I don't make a big deal out of that," he writes in an online bio. "It was and is a very 'so what' phenomenon."

Since that time, he's divided his time between a tree house in Maui and this beach hut in Koh Phangan. He writes books about growing durians, organic gardening and other subjects, including a project on one-word affirmations called *Switchwords*. The idea is that whenever you find yourself in a predicament, you simply need to chant a certain word over and over again and a solution will present itself. For example, if you are hitchhiking and need a lift, simply repeat the word "on." Whenever you lose something, the switchword is "reach." The book is a guide to a hundred condensed incantations.

Nirav's girlfriend, American and in her fifties, makes us banana, cashew and aloe vera shakes with rock dust from ancient seabeds in Utah. "Nirav has been ingesting mineralized rock dust in shakes for years," she says. The two met through Osho. They were both orange-robed disciples of his free-love cult, initially based on a ranch in Oregon called Rajneeshpuram, and then, following Rajneesh's deportation for violating immigration laws, at his ashram in Pune, India. She speaks fondly of their travels in search of perfect mangoes and meditation.

I rent my own beach bungalow, also home to some lizards, which I hear scampering around all night. Before bed, I flip through some guidebooks on fruitarianism. In *Fruits: Best of All Foods,* Klaus Wolfram explains that learning the habit of eating fruits exclusively is an arduous journey that few can complete. Morris Krok's *Fruit: The Food and Medicine for Man* relates how he attended a fruitarian lecture with Essie Honiball, author of *I Live on Fruit.* Asked what one should eat, the answer is: "Fruit, of course." What all the fruitarian gurus have in common is their belief that subsisting on fruits leads to transcendental experiences.

The basic teaching of author Johnny Lovewisdom is that fruits are the way for humans to get back to Paradise, or, as he puts it, "the Hyperborean homeland, the region of sunshine and everlasting spring, where the inhabitants lived on juicy fruits, and knew not what suffering and death were." In his forty-page photocopied treatise, *The Ascensional Science of Spiritualizing Fruitarian Dietetics,* he suggests that fruits can be catalysts for clairvoyancy. "We are not referring to imagining things in the mind," he writes, "but rather a brilliant Technicolor Cinerama experience contemplated by the single eye of the forehead." On a citrus bender in Florida, he witnessed a being descend from the air and materialize out of the ether like Captain Kirk returning to the *Enterprise.* Fruits, he maintains, transform humans into a vaporous essence that can float to heaven like white light. After moving to a crater on the peak of a dormant volcano in the Andes, he spent seven months without eating anything at all and experienced full-blown rapture. Despite his firm belief that fruits bestow eternal life, he died in 2000.

I rise at dawn. Opening the shutters, I see a stray dog walking along the beach and then out into the ocean. The tide is out, but the light bending on the wet sand makes it appear as though the dog is walking on water. He runs out into a mirage that shimmers for miles.

THE SUBSET OF the nutritional demimonde who eat only fruit usually do so by necessity. Druids really only ate acorns and berries. Certain tribes dwelling near the lower Amazon, as well as foragers called *caboclos* often eat nothing but açai berries, brazil nuts and the sap from the milk tree. Nomadic tribes in North Africa wander for long stretches eating little other than dates.

The primary form of sustenance for the hunter-gathering !Kung San of the Kalahari Desert is the mongongo fruit, which has a fleshy exterior and an edible kernel. They also eat !Gwa berries and Tsama melons, which grow underground and contain a liquid that sustains them through periods of drought, but when a !Kung bushman was asked in the 1960s why his people hadn't taken up agriculture, he replied: "Why should we plant when there are so many mongongo nuts in the world?" There was such a plentiful abundance of mongongos in the Kalahari that they couldn't even eat them all. Given the widespread malnutrition today throughout the Sub-Sahel region, I can't help but wonder where the mongongos have gone.

In a famous essay, anthropologist Marshall Sahlins called hunter-gatherers "the original affluent society" because the simplicity of their material wants, needs and diet left them with so much leisure time on their hands. Murders certainly occurred with greater frequency, but hunter-gatherers also enjoyed tight social bonds, and greater support from their families and friends. Mongongos and other foods were always shared between tribespeople. According to anthropologist Lorna J. Marshall, the idea of eating alone and not sharing is inconceivable to the !Kung.

The Seri Indians of Mexico, one of the last hunter-gatherer societies in the Americas, were intimately linked to the pitahaya, or dragon fruit, as it's now known. Food was so scarce that, after eating the cactus fruits in season, they'd rummage through their own feces for pitahaya seeds, which they would then roast and crush for use in the coming winter. As the Seri have been assimilated, younger generations have already forgotten that pitahayas are even edible.

There have been many fruitarians through history. The writings of Buddhist disciples portray Siddhārtha Gautama as a fruitarian. Plutarch wrote that before Lycurgus the ancient Greeks subsisted on fruits. Mohammad is said to have lived on dates and water in Madinah. St. John the Baptist lived on strawberries for a period (some suggest that he actually ate carob beans). Even Idi Amin, the tyrannical Ugandan dictator, lived his final years in Saudi Arabia as a fruitarian (his affinity for oranges earned him the nickname "Dr. Jaffa").

Gandhi's flirtations with fruitarianism were inspired by Louis Kuhne, the frugivorous German author of *New Science of Healing*. Germany was an epicenter of bohemian fruit experimentation from the late nineteenth century until the middle of the twentieth century, when it took off in California. Adolf Just, author of *Return to Nature*, claims that Proto-Germans ate nothing but berries and other forest fruits. The origins of the hippie movement can be traced to German subcultures *wandervogel* (free spirits), and *naturmensch* (nature men). These errant youths congregated at a spiritual community in Ascona, Switzerland, from 1900 to 1920. The rebel health colony, called Monte Verità (Mountain of Truth), espoused fruitarianism and practiced nude organic gardening.

In H. G. Wells's dystopian fantasy *The Time Machine*, humankind bifurcates into two species: the Morlocks, underground toilers who live in darkness; and the Eloi, feeble, childlike little beings who sing and dance in the sunshine and eat only "strange and delightful" fruits such as hypertrophied raspberries and a floury oddity in a three-sided husk.

Although science fiction seems an appropriate realm for this diet, for some it's a real way of life. Fruitarians tell of feeling an ineffable joy on fruits (others might call it fructose-induced delirium). "Many people report telepathy, increased inner awareness, connectedness to all life, a feeling of aliveness like never before, and on and on," writes Rejean "David" Durette, author of the 2004 handbook *Fruit: The Ultimate Diet*. Durette claims that fruits improved his vision to such an extent that, after becoming a fruitarian, he was able to pass his driver's license eye exam without glasses for the first time. Fruitarians "have better looks," says instructor Inez Matus, who says she went from legally blind to 20/20 vision. Female fruitarians talk of their bodies becoming "so chic" that they experience increased male attention. A Japanese fruitarian has reported having such supersensitive hearing that he could make out the sound of ants crawling on the ground or conversations happening six miles away.

Most doctors warn that the fruitarian diet lacks the nutrients required for a balanced lifestyle. There are few objectively validated examples of long-term fruitivores. Another problem fruitarians encounter is a deficiency in B_{12}. One fruitarian, posting on a raw food forum, claims that he gets "B_{12} from performing cunnilingus."

Because it lacks a balanced variety of amino acids, the fruitarian diet is especially dangerous for children. In 2001, Britain's Garebet Manuelyan and his wife, Hasmik, were charged with manslaughter in the malnutrition death of their nine-month-old daughter, Areni, who was fed only fruit. Two years later, Hasmik apparently committed suicide; her body was found in the sea near Brighton. In 1999, Christopher Fink's malnourished two-year-old son was admitted to a Utah hospital by health authorities after it emerged that Fink had been feeding him only watermelon and lettuce. After allegedly kidnapping his hospitalized son, Fink pled guilty to the charges of attempted aggravated assault and second-degree felony child abuse and admitted that the borderline fruitarian diet was the cause of his son's health problems.

Fruitarians seem to prefer the colloquial definition of a fruit rather than the botanical. Some raw foodists I met at a raw food boutique in San Diego (including a young man who was eating only raw cacao nibs and feeling "totally epic") snickered that they had recently busted their fruitarian colleague eating avocados. But isn't an avocado a fruit? Absolutely, say a splinter denomination called rockguacamolians who eat only avocados seasoned with powdered asteroid dust. Cooked-fruitarians claim they can eat everything from pasta marinara minus oregano, soy veggieburgers (hold the lettuce and onions), and peanut butter and jelly sandwiches. Die-hard fruitarians frown on the inclusion of cooked or processed foods.

Some factions eat only fallen fruit. Others refuse to eat any seeds because they contain future plants. One fruitarian Kiawe and Jameson mentioned eats nothing but nonoily fruits and raw pollen. Johnny Lovewisdom, citing Nubian desert hermits eating fifteen figs a day, said only modest amounts should be eaten daily. Others disagree. Dogma abounds. According to the Raw Food Pearamid, fruits should be eaten on an empty stomach. Forget about fruit salads; different kinds of fruit should never be combined. Durette, America's foremost authority on fruitarianism, says the optimum meal should consist of only one kind of fruit eaten to satiety. But with a daily intake of at least ten pounds of fruit, he seems to spend all day eating. Here's an average daily menu:

8:00	1–2 pounds watermelon
9:00	1/2 pound grapes
10:00	1 pound bananas
10:30	1 pound peaches
11:30	2 Hass avocados, 1/2 pound tomatoes
12:30	2 pounds watermelon

1:30	1/2 pound grapes
2:30	1 mango
3:30	1 pound bananas
4:30	3 pounds watermelon
5:30	1/4 pound grapes
6:30	1 pound peaches or 1/2 pint blueberries
7:00	1–2 pounds watermelon
8:00	1 Hass avocado or 1 pound bananas

In the winter, he adds citrus, persimmons and perhaps some almonds and sunflower seeds, although he admits that eating nuts and seeds isn't exactly kosher.

When I contact Durette at his home in Arizona, he says that he's been hiding out to a certain degree (citing an unflattering article in *Playboy*). Nonetheless, he is open to discussing his lifestyle. "The time seems to be right," he says in an e-mail, "for me to get out there and to make my presence known." He explains his belief that protein is overrated, and that there's another sort of unmeasurable life force that fruits bestow. Fruitarians speak of a sense that fruits allow them to tap into some sort of spiritual energy. According to Samuel Riche, a twenty-nine-year-old fruitarian from California, being a fruitarian allows him to attain a state of direct communion with God: "It's almost like living in a realm just outside my physical body—in a lighter plane."

Durette echoes this sentiment, saying that what he eats is a means of getting closer to Paradise. "The Bible makes it clear that we were intended to live in the Garden of Eden and eat fruit year-round," he writes. "We've got to get back to the garden."

THE BIBLE NEVER stipulates that Eve and Adam ate an apple. In fact, the fruit growing on the tree of the knowledge of good and evil is never specified. Apples only started being used to represent the fruit in the fifth century A.D. In Italy, oranges were believed to be the fruits of the tree of knowledge. In 1750, grapefruit was the "forbidden fruit." Linda Pastan writes that it was a pear. Ethnobotanists have proposed the iboga, a West African teardrop-shaped fruit that cures opiate addiction. Others have conjectured that it was a pomegranate, grape, lemon, durian, peach, cherry, coffee berry, pawpaw or even a magic mushroom (a mushroom is, botanically, the fruiting body of a fungus). Enoch, who managed to return to the Garden of Eden (and was subsequently turned into the fiery angel Metatron), reported that the fruit was, in actuality, a date—or, depending on the translation, a species of tamarind.

Perhaps it was a fig. "Truly if I were to say that any fruit had come down from Paradise, I would say it of the fig," declared the prophet Mohammad. The mosaics of San Marco in Venice from the thirteenth century depict figs as well. Even specific varieties were candidates in Midrashic texts: some rabbis said it was a *bart sheba* fig, but another rabbi disagreed, saying it must've been the *bart ali* varietal. Either way, they definitely covered themselves with a fig leaf. Or maybe it was a banana leaf. In eastern Asia, the banana was believed to be the source of Good and Evil. Confusing matters, medieval Europeans called the banana the "fig of Paradise," the "fig of Eve" and the "apple of Paradise." There is still a Ping-Pong–ball-sized variety of banana called the pitogo that looks more like a plump fig than a banana.

Certain sources claim the tree of knowledge contained five hundred thousand varieties of fruit. According to Kabbalistic exegisis, the seven fruits of the Land of Israel—wheat, barley, grapes, figs, pomegranates, olives and date honey—were all taken together in the duo's transgression. In the third century A.D., a gaggle of plenipotent rabbis held a symposium to settle the matter once and for all. As the commentary *Genesis Rabbah* makes clear, in the Jewish tradition, the sin had nothing to do with carnality. One rabbi wonders where Adam might have been while the snake was chatting up Eve. Another rabbi answers: "He had earlier had sexual relations and now he was sleeping it off." The committee concluded that the name of the fruit wasn't specified lest the symbolism of the image be diluted by its connection to something of this world.

As its convoluted name implies, the fruit of the knowledge of good and evil is a metaphor meant to encourage contemplation. It hints at a distinction between the material world and another level of experience beyond duality. The mystical unity of opposites is impossible to grasp with our conscious mind. Fruits were employed, perhaps because they themselves are the coming together of male and female flowers, of sugars and acids, of dying flesh and unborn seeds.

Biblically, the eating of fruits occurs right before humanity is expelled from Edenic bliss into a physical world of good and evil, of unconditional opposites. Twelve more fruits are found on the final page of the Book of

Revelation, all growing on the Tree of Life that grows on both sides of and in the street of a river. These fruits are in season every month of the year. Beyond these fruits, God speaks, saying, "I am Alpha and Omega, the beginning and the end, the first and the last." He is all opposites, unified.

Fruits are symbols that guide us across a threshold into a new reality. Cacao fruit was the "pathway to the gods" in Mesoamerica. In Norse mythology, the Goddess Frigga brought children to paradise bundled in floating strawberries. The hero of *Paradise Regained* is presented with fruits on the doorstep of the kingdom of heaven. This return to the field of creation is called redemption. Examples abound, from the wild fruits outside Eldorado in Voltaire's *Candide* to the apples shimmering like huge soap bubbles in Hans Christian Andersen's *The Bell,* which tells of a young boy's religious awakening. The Gnostics believed the Eden allegory to be, as Elaine Pagels says, "an account of what takes place within a person who is engaged in the process of spiritual self-discovery."

The metaphor is evergreen. In 1982, author William T. Vollmann crossed over the border into Afghanistan with *mujahadeens.* "I will never forget the morning we came to an apricot tree one hill away from a Soviet base. The tree was bowed beneath the weight of its golden fruit. In the sand by the tree was a human jaw." I asked him what it was that affected him about the image. "It was really fertile and normal and gruesome at the same time," Vollmann replied, still overwhelmed by the recollection. It was life and death, or life in death.

Whether it's Aeneas tossing a fruit sop to the three-headed guard dog of Hades or Dante finding a barren fruit tree that bursts to life in between purgatory and heaven, fruits inhabit a liminal boundary zone where they whisper of salvation.

AS A FLOWER DIES, a fruit grows out of it. That brown-gray lint at the bottom of every apple is a desiccated calyx containing traces of a moldy, dead flower. Although fruits ripen and then fall off a branch, they come back the following year on that exact same branch. After a fruit rots and decomposes, the seeds live on. Nature is a feedback loop, from putrefaction to perfection and back again.

Humankind only learned that seeds grow into plants around ten thousand years ago. The miracle of vegetative growth seemed to shed light into the mystery of our own lives. Seeds go in the soil—like dead people.

Perhaps that meant that something equally magical might happen to our own bodies—and souls—after death.

Anthropologist James George Frazer documented anecdotage of tribes around the world who believed their deceased ancestors hovered in fruit trees. *The Golden Bough* mentioned how the Akamba of East Africa believed that "at death the human spirit quits the bodily frame and takes up its abode in a wild fig tree (*mumbo*)." Other hunter-gatherers regarded certain trees as reincarnated fathers. Solomon Islanders told of transmigrating into fruits: "A man of great influence, dying not long before, had forbidden the eating of bananas after his death, saying that he would be in the banana. The older natives still mention his name and say, 'We cannot eat so and so.'"

As I was completing this book, my father told me that he'd been writing his will and pondering his burial. His final wish, he said, is for his ashes to be scattered over his vineyard in Hungary. He then joked that perhaps he'd come back as an indigenous grape varietal: "Oh, this *Badacsonyi szürkebaràt* has quite a distinctive taste—Dad's aftershave."

This idea has long fascinated writers and poets. Gabriel García Márquez once wrote about a woman fearful of eating the oranges in her tree lest it contain remnants of her dead husband. The hero of the ancient epic Welsh poem *Câd Goddeu* evolved from fruits: "Of fruit of fruits of fruit God made me." The Nobel Prize–winning Norwegian author Knut Hamsun once contemplated whether, in a past life, he had been "the kernel in some fruit sent by a Persian trader."

Folk tales all over Europe told of humans descending from fruit trees. In Hesse, a lime tree "provided children for the whole region." In Abruzzi, it was grapes. The Antaifasy of Madagascar believed their common ancestor was the banana. An Indo-Chinese myth describes how the founding mother gave birth to a pumpkin whose seeds became human children. The Sri Lankan myth *Pattinihella* refers to a woman being born in a mango. The Phrygian god Attis was born after his virgin mother wedged a pomegranate between her breasts.

Even stranger than fiction, scientists like Pythagoras and Newton didn't eat certain foods due to their belief in metempsychosis. Pythagoras, like the Chaldeans under whom he studied, thought that souls ended up in fava beans. My college philosophy textbook contained the fragmentary Pythagorean tenet "abstain from beans," not explaining that it might have something to do with eating your forebears. (Heraclitus referenced this in his own aphorism on human fallibility: "Pythagoras may well have been the deepest in his learning of all men. And still he claimed to recollect details of former lives, being in one a cucumber.")

Works of Jewish mysticism, such as the Zohar, characterize human souls as fruits that grow on the Holy One's tree. These soul fruits separate into pairs on their way to Earth. As a result, humans are born with half a soul, a soul that can only be complete by meeting a lover whose soul is the fruit's corresponding other half: a soulfruitmate.

After one visionary experience whilst gazing at a fruit tree, Isaac Luria, founder of Lurianic Kabbalah, told his followers that, "If you had been able to see them, you would have been shocked to see the crowds of spirits in the trees." Other denominations of esoteric Judaism believe that souls, after death, can literally become ensnared in fruit-bearing trees. If someone comes along and makes blessings over the fruit before eating it, the soul is liberated and it enters Paradise. However, if no blessings are made, the soul will be stuck until the end of days. While peeling an orange or slicing into an apple, I find myself uttering gentle words for any disembodied specters.

The tradition of saying grace has been linked to freeing transmigrating souls from their temporal fruit prisons. Buddhist and Jain ascetics eat only fruits that have been cut with a knife or poked by a fingernail because of the spirits believed to reside within. Fijians used to ask coconuts for permission to open them, saying, "May I eat you, my chief?"

AFTER BIDDING FAREWELL to Shunyam Nirav, I catch a flight to Indonesia. Fruits play an important role in the sacred life of Bali, a predominantly Hindu island in an otherwise mainly Islamic archipelago. Shrines abound, plants are dwelling places for supernatural forces and the landscape sparkles with divine potential.

The Balinese live equally in two worlds: *sekala,* the seen physical world; and *niskala,* the invisible spirit world. Fruits, in Balinese ritual, inhabit a zone that is somewhere between the two. Fruit offerings are used to propitiate spirits during ceremonies such as circumcision, marriage, cremation and tooth-filing (a custom believed to eliminate animalistic lusts and desires). Fruits invariably appear at daily contributions to various deities or alongside gamelan musicians playing ancient melodies at rites of passage.

Hinduism was the first major religion to explore the concept of rein-carnation using fruits. The *Brhadaranyaka Upanishad* explains that, after death, smoky human souls waft up to the moon the way a berry detaches from its stem. Once they reach the moon, the souls are eaten by the gods. These souls are reborn when they fall back to Earth in rain and enter plants that bear fruits that are eaten by men that then become semen.

Made "Rico" Raka, the son of a village priest, is guiding me around the island. Some surfer friends of mine have suggested I hire him as a guide, and although he usually spends his time shuttling Australian bodyboard-ers to secluded beaches, his intimate knowledge of the role of fruits in Balinese culture makes him the ideal companion. Because of his reli-gious upbringing, Rico is familiar with temples where little purple fruits with curative properties grow in profusion near shrines crowned with empty thrones. At roadside fruit stalls, Rico explains how various fruits are consecrated to Ganeesha. Bringing me into a coconut grove, he demonstrates how the shell's three eyes represent Shiva. Just as Western-ers break champagne bottles on boat hulls, coconuts are broken to appease the gods at the launch of a ship or at the start of important new endeavors.

We drive out to the country in his van ("Rico's Fun Wagon"), where he takes me into a forest full of ripe *jouat* fruits and snake-fruit trees covered with needles. It's easy to see how they got their name: their scaly brown peel looks like what's left over after a molting snake sheds its skin. At the market in Ubud, we descend into a smoky netherworld of mangosteens, Technicolor pancakes and gelatinous mounds quivering in the half-light. The tiny corridors of Denpassar's *Peken Badung* central market are crowded with women carrying baskets on their heads. Rico sorts through black-jelly nuts, explaining how to eat forest fruits like the *kaliasem, kepundung* or *sawokecik*. When he remembers a special tree near his child-hood home, we drive off in search of it. Pulling to a halt an hour later, we come upon the tree, full of extraordinary pink-fleshed *kinbarans*. Open-ing up their green shells, Rico and I jump for joy.

At the end of our last day together, as we savor some chikus growing in a nearby alley, Rico turns to me and says, "Your name is like a symbol. Think about it: Adam hunting for forbidden fruit!"

While pondering the ins and outs of duality, I go for a walk along the beach as the sun is setting. A priest with a high crown and a golden raiment is conducting a seaside ceremony. A dozen followers chant hymns.

After the sun has dissolved into the roiling ocean, I turn off the beach toward the main drag. Walking through a nondescript beachfront hotel, I arrive at another road I don't recognize. Spotting a little passageway across the street, I follow it, as it winds around through some tree trunks.

A half hour later, lost, I reach a rural area where cars can no longer be heard. I'm on a dusty path, ensconced in a crepuscular silence. Roosters strut about. Looking to orient myself before night falls, I notice the word MELON spray-painted on a barn wall. Some unripe mangoes dangle on a branch just out of reach.

Walking on, I follow a footpath through a rice field. There's a cow flicking its tail in the grass. I wonder if it's sacred. The path brings me to a babbling stream. Venturing over the rickety bridge, I open a wooden gate and enter what appears to be a cemetery or burial ground of some sort.

By now, the moon's out. Shivering, I realize it's become quite dark. The land is desolate. Coconut husks lie rotting in the sand. Heaps of ash suggest recent cremations. A bird of prey circles above. The bushes look like dragons, their heads reared in a silent scream, their talons clawing the air.

It grows colder. And then, two fearsome idols rise out of the dark, their long tongues hanging out, their eyes rolling in their heads. They're holding babies. It's a paradoxical vision of birth in death. I recognize their black-and-white checkered robes, the Balinisian symbol of good and evil. Baskets of fruit lie at their feet. Just then, two security guards shout at me from the other side of the cemetery, and I rush into the glare of their flashlights.

7

The Lady Fruit

And Adam said, This is now bone of my bones, and flesh
of my flesh: she shall be called Woman.

—Genesis, 2:23

M Y LAST DAY in Southeast Asia is spent touring a Rayong fruit
garden filled with Edward Scissorhands–like topiaries. Am,
my lanky, twentysomething guide from the Tourism
Authority of Thailand, is telling me how she used to filch a sour berry
called the *takob* from neighbors' trees. "Stealing fruit so much good," she
says, grinning.

We pause under the shade of a rambutan tree, next to a hedge shaped
like a monkey eating a banana. Drawing a circle around her belly button,
hips and upper thighs, Am asks me if I'm familiar with the "lady fruit." I
shake my head and laugh nervously, wondering if this is some kind of
innuendo. She pats her backside and looks at me questioningly. I freeze.
She keys something into a portable translation device. "In English," she
says, "it's called 'the fruit from which women originate.'"

My blushing subsides as she continues. "People thought it was only a
legend from the Rāmāyana, but now we know the fruit actually exists."
She learned about it from a Buddhist monk who spent many years
abroad living as an ascetic hermit. While wandering through a forest in
India, the monk bumped into a Hindu pilgrim. "In the jungle," Am

explains, "there are places where people with similar beliefs can meet." To commemorate their auspicious convergence, the pilgrim presented the Buddhist monk with a gift: a hardened fruit shell with all the features of a woman's pelvic region, both front and back. Upon his return to Thailand, the monk brandished it as evidence that womankind had indeed evolved from a fruit.

Sensing my doubt, Am suggests that we visit the monastery in a rustic northern Thai village so that I can see it for myself. Unfortunately, my flight leaves the following morning and the monk is a twenty-three-hour train ride away in an area so flooded that people have been sleeping on their roofs for weeks.

BACK IN MONTREAL, I start searching for evidence that corroborates Am's apocryphal story, but literature on the "lady fruit" is scant. According to one seventeenth-century account, a nut fitting the description was believed to grow on an island that could only be found by those not seeking it. Melanesian creation myths explain how the first four men created the first four women by throwing some coconut-like fruits at the ground. I'm convinced that Am's fruit doesn't even exist until I come across a book about magic in India. It mentions that holy *sadhus* in Borgampad carry ritualistic water vessels, called *kamandals,* made from a fruit resembling female butt-cheeks. This fruit, worshipped by a cult, grows only in the Seychelles, where it is called the coco-de-mer.

Armed with a name, I immediately uncover some online images. Not only is the lady fruit real, but it is easily the sexiest fruit in the plant kingdom. Its risqué shell is a life-sized simulacrum of the female reproductive region, including hips, an exposed midriff, two thighs and a pudendal cleft—complete with a tuft of alarmingly lifelike hair on the mons pubis. From the back, it bears a striking resemblance to a woman's derriere. Visitors to the Seychelles call it the pubic fruit, the lewd fruit or the butt nut. Blushing articles in travel sections refer to it as "indecent." "This . . . this is so *dirty!*" gasps a friend when I show her a picture.

Compounding the similarities to human sexual anatomy is this palm tree's flowers. The woody female blossoms are the size and shape of full breasts, with a moist ovule positioned precisely where you'd expect the nipple to be. The male flower, called a catkin, looks like a full-blown erection. When young, the phallic rod is about a foot long, orange and points stiffly upward. At maturity, this arm-sized tumescence becomes spangled with star-shaped yellow blossoms. The pollen from these blossoms fertilizes the female flowers. After pollination, the distended catkin wilts and sags, becoming increasingly brown and shriveled, until it falls to the forest floor with a dank damp thud. The female flowers swell into the coco-de-mer fruits.

As mysterious as its erogenous qualities is the fruit's history. Few even knew the island of Praslin existed until 1756, when European cartographers stumbled onto its sandy beaches. Before then, only some Arab traders, Maldeevian navigators, pirates and rogue seafarers had ever come across the Seychelles Islands, hidden away between East Africa, India and Madagascar. Although uninhabited by humans, the islands' forests were teeming with coco-de-mers.

Prior to the discovery of Praslin, the alluring shell was periodically found floating in the ocean like a wet dream, leading sailors to speculate that it grew underwater (hence the name coco-de-mer, or coconut of the sea). Mariners reported seeing its foliage blowing under the waves. Mermaidlike, it would instantly disappear into the depths, unless sought out by "stout not timorous, religious not superstitious, not weaklings or fools, but judicious and industrious men."

According to Malaysian shadow myths, it sprouted from the central whirlpool from which all life springs. Garcia de Orta, who provided the first detailed description of the fruit in 1563, suggested that it grew on a petrified underwater tree. In Magellan's time, the coco-de-mer was believed to grow in a land called Puzzathar surrounded by maelstroms. Throughout the Middle Ages, coco-de-mers collected by sailors were sold for vast sums. In the seventeenth century, the Holy Roman Emperor Rudolf II bought one for four thousand gold florins. The crowning glory of King Gustavus Adolphus of Sweden's curiosity cabinet was a gold-mounted, coral-sprouting coco-de-mer goblet being hoisted aloft by a silver Neptune. In the Orient, any coco-de-mers found in the wild were automatically considered royal property. Potentates used the fruit in their harems, and Indian religious sects worshipped it in temple rituals.

Before dying in Khartoum at the hands of Sudanese dervishes in 1885, British general Charles Gordon visited the Seychelles and was convinced that he'd found the Garden of Eden. In his manuscript *Eden and Its Two Sacramental Trees,* Gordon created frenzied diagrams attempting to prove that the coco-de-mer was the tree of knowledge of good and evil (the breadfruit being the tree of life).

While flipping through botanical textbooks, I learn that the immature fruit contains a luscious custardlike flesh beneath its salacious exterior. Until the 1970s, distinguished visitors were sometimes honored with a taste of the coco-de-mer's transluscent jelly, then known as the billionaire's fruit. These days, nearly endangered, it is even harder to taste. Since the enactment of a 1978 conservation law, buying or selling a coco-de-mer without a government permit garners a five thousand rupee (eight hundred dollars) fine and two years incarceration. As a result, dozens of coco-de-mer poachers have served—and continue to serve—jail sentences for harvesting the fruit. Lured by the thought of tasting a forbidden fruit, hoping that I am stout and judicious enough for it, I book a flight to the island of Praslin, a speck of equatorial dust somewhere in the Indian Ocean.

PASSING THROUGH CUSTOMS at the Victoria International Airport on Mahé, the biggest island in the Seychelles, I notice a sign asking visitors to inform officials of any fruits or plants they've brought because they can threaten the local ecosystem. Handing them some Canadian Empire apples, my passport is stamped with the shapely outline of a coco-de-mer, the country's official insignia. The airport is filled with Seychellois, the most exotic people imaginable: blue-eyed African princesses speaking French-inflected Creole; young mop-topped descendents of Kenyan slaves working as masseurs; tall, well-endowed Mauritian women with British accents and noses like parrots; and other miscegenated mishmashes of Chinese nomads, Spanish *flâneurs,* Indonesian vagabonds and Sri Lankan seekers.

An hour later, the twelve-seater twin-engine plane to the nearby island of Praslin taxies onto the runway. A landscape bursting with craters and atolls unfolds below. Islets ringed with white sand merge into turquoise translucence as the plane flies low over waves full of leaping fish. Whirling eddies foam at the ocean's surface, as if coco-de-mer trees disappearing into the abyss.

Waiting for a cab outside the Praslin airport, I notice a sculpture of a ten-foot-wide coco-de-mer being spritzed by four bronze-cast fountains in the shape of male flowers. The penislike catkins are ejaculating life force onto an outsized bust of the female genital zone. The thighs, vulva and belly resemble a truncated Paleolithic Venus of Willendorf mother goddess figurine. It's almost pornographic, yet so natural.

The best place to view the fruits in their native habitat is the Vallée de Mai forest reserve. Driving up the steep road to the nature preserve's entrance, I keep catching glimpses of the palms, festooned with creepers. Their trunks are fantastically tall and skinny, stretching over a hundred feet into the sky. The fruits and leaves burst out in a crown at the summit.

It can take a couple of hours to hike through the winding footpaths, and I set off in the company of ranger Exciane Volcere, a jolly, rotund botanist in a khaki uniform. The first thing visitors notice, she says, is the forest's messiness. The ground is strewn with dead branches, leaves, rotting palm fronds, seed husks, plant litter, termite mounds and other natural detritus.

Volcere, who bears a resemblance to Queen Latifah, reaches into a large spiderweb connecting two coco-de-mer palms and plucks out an orange, black and purple spider whose spindly legs barely fit in her two cupped hands. Placing it on her chest, she assures me that it isn't dangerous. It moves rapidly, like licks of flame, over her shirt. To prove its harmlessness, she places the spider on my forearm. It scampers about, surprisingly weightless given its size. It could be eyelashes kissing my skin. As she reaches to pick it back up, it squirts a load of silk at me.

The paths are lined with other local fruiting trees. There's a type of *Pandanus* that bears bristly fruits used as clothes brushes. The large spiky pod from a native spaghetti palm splits open and spills out long strands of al dente spaghetti with capers. The jellyfish tree's stigmas resemble tentacles. The fruit of the kapisen recall the tonsured head of a monk.

These trees are home to a number of endemic fauna: snails the size of grapefruit, rare black parrots and bronze lizards with podlike sucker pad fingers. Volcere points out a fluorescent green gecko licking nectar off a male coco-de-mer flower's yellow anthers. Minute yellow blossoms peek out from between overlapping brown scales on the phallic inflorescence. On another nearby catkin, cell-phone-sized white slugs are also munching on blossoms. It's like watching an educational film about venereal diseases in outer space.

As the geckos' tongues dart out, they flick the flowers, sending microscopic bursts of residual pollen into the air. Pollination occurs through wind dispersal, although legend has it that, at night, the trees move close together and the male and female flowers copulate noisily. Anyone unlucky enough to witness this phenomenon is, it's said, instantly transformed into a black parrot or a coco-de-mer fruit.

After fertilization, it takes seven years for the fruit to mature. When they are very young, the nuts are yellowish and contain a semenlike liquid. A year or so later, at their peak of edibility, the fluid gels to a pudding-like consistency. At this point, the fruit's green outer husk has a thin golden band near the crown. If that golden band is too thick, the interior is too watery. Once this yellow ring is gone, the pinkish-white jelly congeals into a hard vegetable ivory, which lasts until the fruits reach maturity and fall to the ground.

I ask Volcere if it's possible to taste one of the fruits. She says it's strictly prohibited to taste anything in the Vallée de Mai, particularly the coco-de-mers. She does, however, allow me to pick up one of the full-sized shells that has fallen nearby.

The fibrous green husk looks like a pumpkin-sized heart. Having cracked open upon impact, the heart-shaped box comes apart easily. It gives off a pleasant coconutty fragrance. Within this husk is the endocarp (or seed shell) that resembles the female anatomy. The largest seed in the world, coco-de-mers filled with ivory can weigh forty-five pounds. There are documented cases of fruits containing two, and even three seeds, reaching close to a hundred pounds.

If left undisturbed, this seed becomes the beginning of a new tree. After it lands, an eerie cord starts to emerge from the fruit's central slit. The embryonic germ of the new plant is located in the cord's swollen tip. Called a cotyledon, this mutated umbilical cord feeds the baby plant. The cotyledon plunges into the ground, and can then travel up to sixty-five feet away. After burrowing to a spot where it won't have to fight for root space with its parent, this heat-seeking embryonic cable then bursts out of the ground. This first shoot is protected by a pointed sheath so sharp that it can cut through an unwitting foot. Soon after, the first leaves come out, and the tree starts growing skyward.

All that hard ivory within the seed is actually food that nourishes the plant for the first two years of its life, as it makes its way through the understory and up into the light of day. A mature coco-de-mer is a twin tank of fuel that feeds the growing plant. Once the new palm has consumed the ivory, the cord rots and crumbles away, and the empty husk just lies there, having fulfilled its biological obligations. These shells are

sometimes washed into the ocean in heavy rains, and can end up floating around for years. When full of ivory, the coco-de-mers do not float,
so water dispersal is not an actual means of propagating the species.
Even if a full seed were to float to another island, it wouldn't be able to
reproduce because it needs a tree of the opposite sex to pollinate its
flowers.

The life span of coco-de-mer palms remains open to interpretation.
Most people speculate the age to be somewhere between two hundred
and four hundred years, although some suggest that they can live for
eight hundred years. We haven't been around them long enough to
ascertain. Everything we know today about the Seychelles's history
has been deduced since they were discovered in the eighteenth century.

The archipelago's geology, notably the granite boulders jutting out of
the seashore, offer insight into this area's past. These bare rock faces,
mute observers of history's variations, have been dated at over 650 million years old, ranking these Precambrian islands among the world's oldest. This suggests that, until 75 to 65 million years ago, the Seychelles
were part of the enormous megacontinent called Gondwana that linked
South America, Africa, Madagascar and India. After India started drifting
away from Africa, the Seychelles broke off midway and were stranded in
the ocean, where they've stayed till this day, full of life-forms that evolved
in isolation.

Dinosaurs became extinct around the same time the Seychelles got
ditched in the Indian Ocean, leading to speculation that these fruits may
have once been eaten by brontosauruses. "The coco-de-mer could have
been a tasty dessert for a seventy-five-foot-high herbivore," says Volcere,
doing a lumbering dino-dance.

People are afraid of going into the forest after dark, especially when it's
windy. Even in the daylight, the forest is a loud place, with all those
immense leaves clashing against one another and those palms straining
from their loads of heavy nuts in the wind. Things are constantly creaking loudly and snapping. Ancient, guttural gruntings and groanings
pierce the air. It sounds like heavy oak doors being broken apart.

I look up at all the looming coco-de-mer palms surrounding us, their
enormous green hearts swaying ominously above. According to Volcere,
nobody's ever been hit on the head with a falling coco-de-mer. "But if

one day it happens, they will surely stay here in paradise," she says, laughing morbidly. "It's a good souvenir for them, no?"

The Seychellois have a complicated relationship with the tourism industry that both keeps this economy afloat and also threatens to destroy its ecology. I ask if Volcere's ever spent the night in the Vallée de Mai. Yes, she says. Last night in fact.

"Why were you here?" I ask.

"Did you hear about the kerfuffle with the opposition party?"

I had, earlier that morning, read in the newspaper about how the police had battered the opposition leader at a demonstration outside parliament over the right to start their own radio station. (There is only one radio station, owned and operated by the state.) As tensions mounted, the head of police pistol-whipped the opposition leader, who required twenty-six stitches to the back of his head.

"Last night, the army was deployed to guard the Vallée de Mai because the opposition followers threatened to burn it down. I spent the night here with them."

"Why would they burn down the forest?"

"Supposedly to protest the beating, but also as a way of eradicating tourism," she says, unsure whether she should even be discussing this. Freedom of speech is still approached with uncertainty in this fledgling democracy. According to Volcere, opposition supporters say the threat of burning the forest is just a smear tactic by the government and they would never dream of destroying their arboreal heritage. An accidental 1990 forest fire in nearby Fond Ferdinand wiped out wide swaths of forest that will take hundreds of years to recover. It's staggering to think that something so ancient and precious is also so precarious.

THE ARE ONLY 24,457 coco-de-mer palms left. Two-thirds of them are too young to bear fruit and half are male trees. The 2005 census revealed that only 1,769 fruits actually reach maturity each year—not many when you consider that close to one hundred thousand tourists visit annually. The commerce must be regulated to ensure that coco-de-mers won't be overharvested, says Lindsay Chong-Seng, director of the Seychelles Island Foundation, an environmental group that oversees the Vallée de Mai. "There's a lot of poaching going on," he explains. "Thieves even cut entire trees down to get the nuts. They take off on fishing ships in the middle of the night."

Although extinction concerns have classified the fruit as vulnerable on the World Conservation Union's Red List, management efforts have been successful so far. The entire neighboring island of Curieuse, where the palm also grows, has been cordoned off as a nature reserve. The government maintains a database listing every coco-de-mer tree in the Seychelles, and it is the owners' responsibility to submit—under penalty of law—quarterly statements about each fruit's level of maturity.

After the fruits fall, their ivory is extracted. The hollow shells are then issued permits and sold to tourists at licensed outlets for two hundred to a thousand dollars. Anyone attempting to export an unlicensed coco-de-mer has it confiscated and is fined. There are loopholes in the legislation, including forged permit tags and repeated use of the same tags, which are flimsy green stickers stamped with an official seal. "There should be proper certification," says Chong-Seng, "perhaps a computerized system, or microchips."

New avenues of coco-de-mer conservation are also being explored, including selling fiberglass nuts to tourists. The prototypes I saw were utterly realistic and lifelike. Another idea being proposed is that rather than merely bringing home an empty shell, visitors should be offered the option of buying a mature seed full of ivory and then planting it. Once the plant grows up, there would be a plaque with the donor's name next to the tree, and the emptied husk would be held for them until they next return. The concept reminds me of Ken Love's Hawaiian lychee-planting project.

I ask Chong-Seng whether there is any way to taste the fruit. Shrugging, he says there is only one legal recourse. "If I myself had a coco-de-mer palm on my property with immature fruit ready to be eaten—which I don't—I could invite you to taste it, but I wouldn't be allowed to sell it." In other words, buying it may be illegal, but sharing it isn't. "If you can find someone with a tree in their backyard, all your dreams will come true," says Chong-Seng. As we say good-bye, he gives me a tip: "Do what you'd do anywhere else—speak to the taxi drivers."

EATING AN IMMATURE fruit involves an ethical consideration: it will not become a new tree. Chong-Seng has assured me that isn't necessarily a problem—none of the mature fruits sold to tourists are used to grow new trees either. The vegetable ivory that would otherwise be nourishing young plants is removed by a company called Island Scent, Ltd.

At their small factory in Mahé, I watch employees chisel pieces of coco-de-mer flesh out of the husk. These chunks are dried and packaged, and then maintained in a temperature-controlled chamber before being shipped off to the Far East. In China and Hong Kong, coco-de-mer slices sell for 130 dollars a kilo in herbal shops alongside tiger bones, flying lizards and powdered rhinoceros tusk. The flesh serves many functions: in Malaysia, it is blended into face-whitening creams; in Pakistan, it is used as an aphrodesiac; in Indonesia, it is added to cough syrups. One Middle Eastern businessman pays Island Scent to ship loads of shells to El Paso, Texas, where they are then smuggled underground into Mexico to be inlaid with Arabic designs, motifs and letters. Called *kashkuls,* they are sold to homeowners and mosques in Kuwait and Iran.

"Coco-de-mers have long been used in rituals by Persian fakirs and Indian mendicants," explains Kantilal Jivan Shah. Kanti, as he is known, is an octogenarian Seychellois historian, and environmentalist. I've come to see him because he was instrumental in persuading UNESCO to designate the Vallée de Mai as a World Heritage Site. The owner of an old knickknack emporium that sells fabric by the yard, Kanti is somewhat of an authority on the fruit. "It's used in so many bloody ways," he tells me, his tired eyes sparkling and his smile shining with gold teeth.

Renowned throughout the Seychelles as a palm reader, Kanti's quite the multihyphenate, as I learn when he hands me his CV. "I'm a guru, I'm a cook, I'm a sculptor and I'm also a shah," he says with a strong Indian accent. His sparse, wiry white hairs flutter in the breeze, giving him a hint of amiable lunacy. "I'm a priest. I'm a stamp designer. I'm a healing medicine man. I do many things."

Moments after I arrive, an Italian tourist pops into Kanti's shop. They start talking about traditional Creole architecture. "I'm elected to the International Society of Architects," Kanti points out, winking. An Iranian honeymooner leafs through Kanti's photo album and asks how he happened to invite Empress Farah Pahlavi over to his house for dinner. "She invited herself because I'm a hell of a guy," he replies. "I used to throw big parties for all the top guys in the world."

A self-proclaimed master in the art of harnessing the energy of crystals, Kanti says he is also a respected numismatist (a student of money), chromotherapist (a color healer) and conchologist (a shell collector). "I work in mother-of-pearl like nobody. I design according to the zodiac. I'm a fellow of the Royal Geographic Society. I'm a Jain. I never go to a restaurant."

"You sure seem busy," says the Italian architecture student.

I try to steer the conversation back to the coco-de-mer. Kanti says he's written about the fruit, but gives up on finding the article after pushing around a few stacks of yellowing papers. "It's such a mess here, I can't find anything," he says. "I do too many bloody things. I'm on every damn committee. I'm the treasurer of the alliance française."

"Look, Hindu holy men have used it for centuries as a begging bowl," he explains, flipping through a scrapbook of his exploits, which include lending swords to Roman Polanski, acting opposite Omar Sharif and inspiring Ian Fleming's character Mr. Abendana in *For Your Eyes Only*. "The fruit is a symbol in tantra, a devotional cult object. It is venerated as the yoni, a symbol of creation and fertility."

Although the coco-de-mer still has all sorts of medicinal and mystical properties attached to it today, Kanti dismisses claims for its supposed aphrodisiacal properties. "It's all in the mind," he scoffs. "The dried kernel irritates your bladder so you'll have an erection. You know what it's like at four in the morning when your bladder is full? It's the same bloody feeling."

LEAVING JIVAN IMPORTS, I head to a nearby gift shop to pick up my own sustainably harvested fruit shell. The boutique's back wall is stacked to the ceiling with coco-de-mers. Some are huge, some are small. No two are alike. I pick up ones that are rounder, slimmer, more bulbous, flatter. It seems impossible to chose. "What do you like in a woman?" asks the shopkeeper, fondling a svelte Kate Moss-like beauty. "That's the way to pick a coco-de-mer. I personally like slender women, so I would choose this one."

As the shopkeeper hands me the proper certificates and permits, I ask him if he's ever tasted the fruit. Yes, he says, several times, when he was younger. "It's a taste of wilderness," he says. "It's out of the ordinary, kind of like a minty sperm."

Remembering Chong-Seng's advice, I'm hailing cabs as often as possible. The first driver can't help. In fact, he recently retired as a policeman, and has been involved in sting operations to capture coco-de-mer poachers. Another driver offers to climb a palm and harvest one for five hundred dollars. Although tempting, the whole scheme is beyond my moral, legal and financial comfort zone. Another suggests that I head to the beach, where "touts and beach boys" sometimes sell stolen coco-de-mers by the spoonful. The lifeguard on duty, however, informs me that the practice tapered off years ago. "They'd be better off carrying a bag of dope," he says. "They wouldn't last two seconds."

On the way back from the beach, my taxi driver is a Rastafarian who says he eats coco-de-mers all the time. I ask him how they taste. He pauses.

"Have you ever tasted breast milk?" he asks.

"Um . . . not since I was a child," I answer.

"Well, the taste of the fruit is quite raw," he continues. "It's very . . . *personal*. It tastes like mother's milk straight from the breast."

The Rasta agrees to meet me later that afternoon to discuss procuring a taste of the coco-de-mer. I meet him at a bar overlooking a white sand ocean as the sun is setting. The water looks like molten golden butter. Large crabs scamper up regular coconut trees, cut the nuts off with their pincers, and whisk back down to eat the fallen fruits. Fruit bats the size of seagulls circle close above.

At first, he's adamant that I won't be able to taste it. "We'd both be shot," he says. "It's totally illegal." I point out that paying for it may be a crime, but all we need to do is find someone willing to share it with me, rather than sell it to me. After a few drinks, he softens: "Well, it can always be tried." He starts making some phone calls, speaking a rapid Creole. Hanging up, he smiles widely. "I'm 80.56 percent sure we can get some."

AS MUCH AS I've learned about the lady fruit, it's still as enigmatic as it was when Am first described it to me. That ponderous, subpendulous male catkin. Those female breast flowers that ooze nectar at the nipple. The heart-shaped exterior. The bum-shaped nut, which resembles kidneys when opened. The vaginal umbilical cord. Its genetic similarities to humans seem so overt, yet perhaps it's merely an elaborate evolutionary coincidence. Although it may not be the fruit from which women originate, certain parallels may emerge once its DNA is mapped. All species share a common ancestor, so the possibility of an overlap isn't totally preposterous. After all, the same substance—lutein—found in human retinas, is also found in green plants.

While awaiting word from the Rasta, my final afternoon in Praslin is spent with John Cruise-Wilkins, a fifty-year-old history teacher who believes he's close to finding a treasure buried by the pirate Olivier Levasseur in the eighteenth century.

Just before being hung from a tamarind tree on Réunion Island in 1730, Levasseur flung a parchment bearing a cryptogram into the crowd of onlookers. In 1949, this coded message came into Cruise-Wilkins's father's possession. He spent sixteen-hour days trying to decipher it, and finally came to believe that the note contained directions on performing a series of tasks influenced by the twelve labors of Hercules. Completing these tasks would lead to discovering further clues. Over the last twenty-seven years of his life, his father excavated extensively, and unsuccessfully, in the beach across from his home, using custom-made pumping machinery now rusting nearby.

John Cruise-Wilkins shows me the clues they've found so far: a rock that sort of looks like a woman's head that supposedly represents a waterlogged sarcophagus of Aphrodite's torso at low tide, a rusting horn symbolizing the cornucopia, shards of pottery marked with something resembling a fleece, and a stone shaped like a sandal ("It's the sandal of Jason!"). As he pokes through cabinets in his ramshackle bungalow, a red bird flies in, flutters around freely and then zips back out the window. His father was a big-game hunter before getting treasure fever, and the walls are decorated with antlers, skulls and other trophies. The family's most persuasive clue is a boulder in the shallow tide with a slit carved down the middle. Leading me out to the beach, John says it represents a key. To me, it resembles a giant, granite coco-de-mer.

Burly and intense, Cruise-Wilkins has the cloudy, pale blue eyes of a believer. As he reveals a particularly convoluted clue—a piece of rock that resembles the skeleton of Pegasus—the left side of his mouth curls up ever so softly. His cheeks lift as though pulled up by some invisible pulley system, and the wrinkles streaming from his eyes tighten. His face brightens for just an instant! And then . . . Then, the eyebrows descend like thunder as the inescapable realization crushes him yet again, the same relentless search that haunted his father now tormenting him.

"We're on the verge of discovering the treasure—as you can see, the evidence is overwhelming," he says, picking up a piece of fossilized coral in the shape of the letter Y. Everything seems impregnated with significance. As we head back toward his home, a vivid rainbow spans the sky, landing on a red house in the hills across the bay. "Yes," says Wilkins. "It's a good sign of the covenant between Noah and God." We stroll on in silence for another moment, ensconced in the human need to find patterns, to make sense of all the fragmentary details.

He stops on the side of the road. "People in authority say 'You're a dreamer, Wilkins.' That may be, but I'm a practical dreamer with my feet on the ground. People tease me. They mock me, but I know this is true. I'll prove my father was right."

As we leave, my taxi driver shakes his hands and says, "Never give up hope." Cruise-Wilkins's eyes flash. "It's not about hope," he says acidly. "It's about reality based on historical archeological evidence."

BACK IN MY HOTEL room at the Lemuria, a luxury resort that is a far cry from Nirav's Thai beach hut, the phone rings. The Rastafarian taxi driver tells me that Les Rochers restaurant has some coco-de-mer palms in the yard. I'm to show up there for dinner and "everything will be arranged." Although somewhat dubious legally, it seems like a fair exchange—I buy dinner, they share a taste of the fruit. I'll finally be able to sample a coco-de-mer. Throwing on a raincoat, I race out into heavy showers.

By the time I arrive at the beachfront restaurant, the downpour has tapered off. Unfortunately, the owner explains that he hasn't been able to pick a coco-de-mer because of all the rain. "It's impossible to send anyone up the palm because when it's wet it becomes very slippery and dangerous," he says apologetically.

As I eat, it hits me that I've come all the way here to the other side of the world and I haven't managed to taste the elusive billionaire's fruit. Perhaps it was an unreasonable expectation. But as I finish up my octopus curry, I start wondering whether any other restaurants might have coco-de-mer palms. I think back to a lunch spot, Bon Bon Plume, where I noticed a grove of palm trees in the backyard.

As soon as I get back to my hotel, I make a call. The restaurant owner, Richelieu Verlaque, picks up. Explaining that it's my last night in the

Seychelles, I ask if he might be interested in offering a journalist a sample of coco-de-mer. Verlaque says that he's been planning to eat one the following morning and that he'd be happy to give me a taste. He suggests I arrive at 6:30 A.M. so that I'll be able to make my departing flight at 9:30 A.M.

"Are you sure it isn't illegal?" I ask.

"I can do whatever I want on my property," he booms. "And if I want to give you a taste of coco-de-mer that's my business."

"What about all the rain?"

"That isn't a problem over here. The fruits grow on a dwarf coco-de-mer tree. To harvest one I just reach out and cut it off. In fact, there is a fruit that is perfectly ripe right now."

I LEAVE MY HOTEL before dawn. After driving around the darkened island for forty-five minutes, I arrive at Bon Bon Plume shortly after sunrise. It's already scorchingly hot. Richelieu Verlaque greets me at the door, and we walk out to the backyard together. We pass a giant tortoise in a sandbox next to the restaurant. The tortoise moves very slowly, but cranes its serpentlike head out of its shell when I crouch down to greet it. It looks up at me with tears in its eyes.

Verlaque ushers me toward a picnic table, sweeping his arm over a platter of sliced lady fruit. I pick up a piece, noting the thin golden band indicating ripeness. The fruit's innards are transluscent, almost like a silicone gel implant but with a softer, shaky-pudding texture more akin to a real breast. I bite into the gelatinous flesh. It has a mild citruslike quality, refreshing and sweet with earthy, spunky notes. It tastes like coconut flesh, only sexier.

"So you are one of the rare few who has tasted the forbidden fruit," declares Verlaque triumphantly. "Adam has tasted Eve."

Sitting at the table, we chat about the political climate, and what the assault on the opposition leader could mean for the country's future. We eat some more coco-de-mer, cool in the heat of this sunny Sunday morning. Henry André, Verlaque's eight-year-old son, comes over and eats a few slices with us.

"What do you think it tastes like?" I ask him.

He replies immediately, as though it were obvious: "It tastes like coco-de-mer!"

8

Seedy: The Fruitleggers

Crates of melons on sidewalks, bananas coming off
elevators, tarantulas suffocating in the new crazy air,
chipped ice in the cool interior snow of grape tanks . . .
All of it insane, sad, sweeter than the love of mothers,
yet harsher than the murder of fathers.

—Jack Kerouac, *Jazz of the Beat Generation*

P ERHAPS the only thing more complicated than tasting a coco-de-
mer is bringing one home. Customs and immigration forms
require that all fruits, food, plants or plant parts be declared.
Importing an endangered species, even a sustainably harvested one, is a
process fraught with preemptory guilt. As the plane descends into Montreal's airport, a bubble-butt of a coco-de-mer throbbing in my checked
luggage, I take a deep breath, trying to steady my mildly irregular heartbeat.

Staring at the declaration form, I consider lying. In the event of a
search, I reason, I could try to convince them that it's a sculpture—some
sort of exotic erotic folk craft. I then realize that the permits would incul-
pate me were the nut to be unwrapped. So I check the box, still uncertain
what to say.

Arriving at the interrogation booth, the customs agent glances at my
form and immediately asks what foods or plants I'm bringing in. I start

listing the goods I picked up on my stopover in Paris: apples, dried apricots, a couple of bottles of wine.

The agent writes them down: "Anything else?"

"Well, I have . . .

This is it—I'm one wrong move away from a cavity search.

. . . some nuts."

That's it—NUTS! Oh, sweet nuts of truth. It comes in a flash, but it's technically accurate. After all, I have some trail mix in my backpack, and that unholy coco-de-mer—why, it's just another nut, legitimately declared, so what seems to be the problem, Officer? The agent scribbles the word "nuts" at the bottom of the grocery list, adds some secret codes and waves me on.

Pulling my suitcase from the carousel, I'm feeling poised, even though I haven't broken on through to the other side quite yet. I'm merely in a twilight zone of duty-freedom. I still have to hand my declaration form to the next level of agents who will instantly recognize the ciphers scrawled over it in red ink.

When my turn comes, the bored officer holds his hand out. High on the brazenness of nuts, I disdainfully submit my form and breeze past him. I'm about eight seconds away from the exit. I don't hear any protestations, so I keep on going. "Free," trills my heart, "I'm free!'

Six seconds to go.

And then: "Mister? Excuse me, mister!"

I keep on walking, pretending not to hear. Without wavering in velocity, I attempt to convey the countenance of a corporate type with massive fiscal responsibilities.

Four seconds.

"EXCUSE ME," comes the voice, again. I continue forward. A muscle spasm ripples across my back. A bead of sweat appears at my temple and squiggles away.

Two seconds.

The sliding doors open.

Footsteps pound behind me on the terrazzo. "Stop right there!" yells the officer.

"Who, me?" I ask, craning my neck around and feigning befuddlement.

"Yes, you," he replies sternly. "You can't bring any apples into the country. Step this way."

I start rehearsing as I head toward the inquisition zone. "Well, actually, I did declare all my nuts, including that one . . . Anyways, it's just a sculpture." My blood is so overheated that I can feel the wax melting in my ear canals.

Remembering that the customs official had mentioned apples, I take the sack of French heirloom apples out of my backpack. With trembling hands, the first thing I do is hand them over to the agent sitting at a polished aluminum counter, hoping to distract him with my openness. "I was told you were going to confiscate my apples," I say jauntily, but with a hint of irritation (the better to confound him with my naturalness). Does he realize I'm hyperventilating? The agent takes the apples and, stifling a yawn, points at a doorway. What's this? The frisking chamber? I push open the door, and find myself in the arrivals area, surrounded by smiling families milling about and greeting their loved ones.

IN AUGUST 2005, fifty-seven-year-old farmer Nagatoshi Morimoto was convicted of smuggling 450 illegal citrus cuttings into California from Japan. Boasting to associates that "nobody's gonna catch it," he placed the bud wood in boxes marked as containing candy and chocolate. He was wrong: the stash was intercepted at customs. Under the Plant Protection Act, Morimoto was served with a five-thousand-dollar fine and sentenced to thirty days in jail. He also had to perform community service—handing out brochures informing farmers about the dangers of smuggling.

Because fruit-fly infestations are a major threat to agricultural systems, it's imperative that certain fruits be kept out. Morimoto's "candies and chocolate" contained citrus canker, a disease that could cost California up to 890 million dollars in the event of an outbreak. A medfly plague would be even worse: if Mediterranean fruit flies were to take hold in California and continental America, it is estimated that they would cause 1.5 billion dollars in damages annually. No wonder border guards are so stringent about screening fruits.

Periodic infestations of medflies cripple harvests, forcing growers to shut down production as inspectors seek to eliminate the insects. Medflies, which lay their eggs under the rind of ripening fruit, cause spoilage, deformities, lowered crop yields and premature fruit drops. They also make fruits subject to export sanctions. Their ongoing presence in Hawaii has wreaked havoc on the state's fruit industry.

Pest concerns are taken seriously. USDA officials conducting a routine search of *Dawn Princess,* a cruise ship docked at the port of Los Angeles, recently found dozens of mangoes swarming with live fruit flies. In the early 1990s, two air cargo containers full of illegal Thai fruits covered in pests were seized in California with an estimated street value of 250,000 dollars. When that border agent asks you to hand over your oranges, he or she won't be eating them. Seized goods are ground up, trashed, burned, autoclaved, buried in landfills or whisked away to oblivion by haulers licensed to transport medical waste.

Because of the hazards involved, it's guilty until proven innocent in the produce world. Substantiating a fruit's safety is a costly process that requires years of research. Smuggling fills the void. Nobody knows exactly how much the global fruit black market generates, but the worldwide underground trade in protected flora and fauna generates an estimated 6 to 10 billion dollars annually.

Without fruit flies, importing fruits would be easier. Even though free trade agreements have eliminated many tariffs and other trade obstacles, unfounded phytosanitary concerns are often cited as a carte blanche to bar the import of foreign products. For cases when fruits genuinely harbor pests that could endanger domestic crops, these measures are vitally important. But in countless other instances, they are merely a way of blocking import from developing nations.

The Millennium Development Goals of the World Trade Organization are intended to allow an easier flow of goods into developed nations. The WTO is finally catching on to the false sanitary exemptions, and has started enforcing a more scientific approach to nontariff barriers. Their overall objective is "to permit countries to take 'legitimate' measures to protect the life and health of their consumers (in relation to food safety matters), while prohibiting them from using those measures in a way that unjustifiably restricts trade."

A modern-day example of the complexities of getting fruits into North America and Europe can be seen in the saga of the Indian mango. There are over 1,100 varieties of mangoes. Some are the size of Ping-Pong balls, and others weigh over five pounds. All we usually get is a variety called Tommy Atkins, an early twentieth-century military term for a faceless soldier. It's a fitting designation. Tommy Atkins mangoes are rugged, robust and fibrous warriors fit for the rigors of international commerce. They barely resemble delicious South Asian mango cultivars like the *madhuduta* (messenger of fragrance), *kamang* (embodiment of Cupid), *kokilavasa* (abode of cuckoos) and *kamavallabha* (the amorous). The most popular variety, the Alphonso, is the opposite of the Tommy Atkins: it boasts complex flavors, is fully fiberless and pools nectar to its surface when bitten. Peeled like an orange, it is eaten out of hand like a messy apple. For connoisseurs, the best part of eating an Alphonso is licking the juice off their fingers, hands and arms.

Indian mangoes weren't allowed into the United States for nearly thirty years, ostensibly due to pest concerns. The real reason was nuclear. India and Canada had signed treaties that provided India with CANDU nuclear reactors intended for civilian use. In 1974, it was discovered that the Canadian reactors were covertly being used to manufacture plutonium and build a nuclear arsenal. After breaking off their relations due to nonproliferation violations, Canada and India resumed nuclear trade in 1989—the year mangoes stopped coming into the United States. Every spring, crates of Alphonsos pour into Canada on British Airways cargo planes. I ate four boxes (forty-eight mangoes) in 2006. But I couldn't have eaten any in the United States.

That changed in 2007, following India's new nuclear treaty with the United States. When Indian commerce minister Kamal Nath met with U.S. Trade Representative Rob Portman to discuss a multibillion-dollar deal for the sale of civilian nuclear technology, mangoes were on the table. Nath assured Portman that the nuclear deal would be green-lit—on the condition that India's mangoes be allowed back in. A few months later, President Bush flew into India to discuss the deal, announcing "the U.S. is looking forward to eating Indian mangoes."

Mango diplomacy, as the media have dubbed this phenomenon, demonstrates that the machinations of trade are a form of geopolitical tic-tac-toe. Any small grower trying to import pincushion fruits or ice cream beans would be mummified in red tape without well-placed insiders. Shipping fruits in accordance with phytosanitary regulations involves years of technical procedures and tests. These delays, which in the past were interminable, and hence another masquerade for protectionism, are now being monitored as part of the WTO's imposed measures to limit nontariff barriers. This process will continue to require administrative delays for pest risk analysis and regulatory reviews, but in coming years, it will be harder to cite unwarranted phytosanitary concerns as a means of keeping out third-world imports, resulting in a greater assortment of exotic fruits on supermarket shelves.

In the United States, these fruits will all be irradiated before entering the country. This requires that the exporting countries invest millions in inspection and irradiation facilities. Until recently, nuclear irradiation chambers were used to beam fruits and other foods with radioactive cobalt isotopes. Because of the consumer outcry (owners of fruit X-raying facilities were receiving death threats), a new technology has now been adopted called "electronic pasteurization." These irradiation units are powered by electron beams derived from electricity, rather than nuclear by-products. It appears to be safe for human consumption, but opponents claim it is irradiation just dressed up: the chamber still pelts fruits with electrons traveling at the speed of light. Nonetheless, the process has rapidly been adopted all over the world. Brazil, touting itself as "the fruit basket of the world," has built dozens of facilities. Electronic pasteurization prolongs shelf life and destroys microorganisms and small insects, while supposedly only slightly altering the fruits' basic nutritional integrity.

Until this technology was given the go-ahead in Hawaii in 2000, the rambutan wasn't allowed into the United States. When it first arrived in New York City, entire shipments sold out within hours at high-end grocery stores. No one seemed to know—or care—that they had been electronically pasteurized. Even if they were labeled, the choice is between zapped rambutans or none at all.

Another method is hot-water baths. Some exotic fruits imported into the United States spend four hours in saturated water vapor at 117.5 degrees Fahrenheit. Boiled, underripe mangoes: yum!

Even after all the preventative measures, some fruits still harbor larvae. They are sent to fumigating chambers before being released to the wholesaler (who isn't required to announce the fumigation). And because snakes, spiders and other beasties occasionally slither in with fruit shipments, holds are often gassed before unloading.

Many Pacific Rim fruits cross into the United States from British Columbia, where there is no fruit-fly threat to Canada's temperate crops. The demand for forbidden fruits ensures a high price, so the risks of smuggling them into the States can be seen as worthwhile, especially for shady produce sellers catering to newly emigrated customers desperate for a taste of home. In 1999, Tu Chin Lin spent five months in jail and five months in home confinement for smuggling contraband longans into Manhattan's Chinatown. He was part of a fruit-rustling organization that obtained illicit fruit in Canada and smuggled truckloads of booty over the border by falsifying invoices and customs documents.

The mangosteen, during its years on the lam, could occasionally be found in U.S. Chinatowns. When I'd unwittingly smuggled mangosteens to Ossenfort in Manhattan, we scoffed at the idea of fruit smugglers. Little did we know the phenomenon is so widespread that Nintendo recently released a fruit-smuggling video game called *Bangai-O*. Its premise is that the SF Kosmo gang has hijacked a shipment of intergalactic fruit, which they are selling at inflated prices. The player's aim is to destroy this fearsome network of fruit pirates.

Some countries have ever fiercer antismuggling laws. New Zealand stringently monitors fruit imports, as rockers Franz Ferdinand and actress Hilary Swank learned when, on separate occasions, they were fined for not declaring their apples and oranges. Thirty-four-year-old Chinese student Jian Lin was nabbed smuggling five mangoes and fifteen pounds of lychees into New Zealand. The fruits, which were not declared, showed up in an X-ray inspection of her luggage. Lin pleaded guilty to breaching the Biosecurity Act and was fined 1,000 dollars. The judge was lenient; the fine could have been 150,000 dollars.

In 2004, a Japanese tour guide was fined thousands of dollars for smuggling eleven pounds of peaches into Australia. Other countries face distributors who smuggle as a means of bypassing tax charges. Syrian and Jordanian fruits are smuggled through Lebanese customs so they can be sold duty-free in markets. Rampant smuggling of fruits between China and the Association of Southeast Asian Nations has been countered with a zero-tariff free trade agreement. In July 2005, the *Bangladesh Independent* reported that vast quantities of mangoes, apples and grapes were being smuggled in by corrupt Bangladesh Rifles, a paramilitary force.

In the United States, stiff new penalties are acting as a deterrent. The Fruit, Vegetable and Plant Smuggling Act of 2001 has resulted in felony sentences of up to five years incarceration and fines of 25,000 dollars. Repeat violators get up to ten years of imprisonment plus another fifty-thousand-dollar fine. Even minor smuggling acts are misdemeanor crimes punishable by a year in prison with fines of a thousand dollars and more for repeat offenders.

Officials have started paying informers a percentage of the fine levied. Concerned parties can call a hotline if they have information about the smuggling of prohibited exotic fruits. In the United States, a number of different federal agents chase smugglers. Customs and Border Protection (CBP) is the jurisdiction of Homeland Security. The U.S. Department of Agriculture oversees an organization called Plant Protection and Quarantine (PPQ), as well as the Animal and Plant Health Inspection Service (APHIS).

APHIS, in turn, has set up an antismuggling unit called the Smuggling, Interdiction and Trade Compliance (SITC). Its hundred employees, with an annual budget of 9 million dollars, conduct intensified cargo blitzes at U.S. ports every couple of weeks. Over sixty-eight tons of prohibited Asian fruit were seized in the organization's first two years. Special agents at the U.S. Fish and Wildlife Service recently conducted a sting called Operation Botany that nabbed a ring of plant smugglers. Regional subdivisions include Closing the Los Angeles Marketplace Pathway (CLAMP) and Florida Interdiction and Smuggling Team (FIST). The FIST comes down hard, raiding suspicious greenhouses (like Richard Wilson's Excalibur Nurseries) with attack dogs and machine guns.

The process of importing and exporting fruits is an unceasing stream of forms, e-mails, faxes and other flotsam. This paperwork, which has increased greatly since 9/11, ensures that any fruit crossing a border can be traceable to its warehouse of origin, in case of any terrorist maneuver involving produce. The computerization of the shipping infrastructure has resulted in cargo being declared before trucks reach border checkpoints. New breeds of scanners have been developed to identify fruits and plants in baggage. The Beagle Brigade, a doggy inspection program manned by beagles with names like Liberty, patrols many U.S. airports.

Another reason for the draconian crackdown on fruit smuggling in recent years is that massive quantities of drugs come into North America inside fruit shipments. In 1990, one inspector verified a shipment of 1,190 boxes of canned passion fruit; a tenth of them were full of narcotics. Notorious Colombian drug smuggler Alberto Orlandez-Gamboa shipped cocaine into New York inside banana skins. The Mexican drug cartel headed by Amado Carrillo Fuentes brought in tons of drugs each month on fruit-carrying eighteen-wheelers and 727 airplanes (for which Fuentes became dubbed the "Lord of the Sky"). In November 2004, a shipment of Hit Fruit Drink cartons containing 1.7 million dollars' worth of liquefied heroin was seized in Miami. A disheartened health-food store employee told me that his rich bosses had been moving ayahuasca and cocaine out of the jungle in shipments of dried fruits. Convicted smuggler Richard Stratton brought fifteen tons of Middle Eastern hash into America inside cartons of dates.

An Australian-Colombian banana importer was arrested in 2003 after 35 million dollars' worth of cocaine was found stashed in banana crates. Officers raiding his office at Kristel Foods found another 9 million dollars in cash. Seven Chiquita banana ships were stopped in 1997 containing more than a ton of cocaine. Another fruit industry professional in Montreal told me that almost all the city's drugs arrive sewn up in fruits. Even though fruits are tracked and documented at every transfer point, it remains impossible to search more than a small fraction of the cargo. "You think they're going to open every box of mangoes at every border?" he said. "Anyways, everybody's getting greased. Customs guys get thirty thousand dollars in brown envelopes to let a truck through. That's how much they make a year. You think anybody's saying anything? If someone does get nailed, they'll blame the guy at the loading docks. They'll say his cousin in Honduras sent it."

In the summer of 2007, police uncovered 38 million dollars' worth of cocaine at the port of Montreal. When I heard the report on the radio, I turned up the volume, certain that fruits were implicated. Sure enough, the drugs were found in buckets of frozen mango pulp.

Fruit trucks are also used to smuggle human immigrants. In 2007, immigration agents in Huixtla, Mexico, were tipped off by the smell of human sweat when searching an eighteen-wheeler full of bananas. They found ninety-four people hiding among the fruit crates. The bust revealed that Carlos César Ferrera, "King of the Trailers," had been overseeing a network of hundreds of trucks carrying human cargo. Ferrera would approach truckers and ask whether, for a fee of five thousand to ten thousand dollars they'd be willing to carry what he called a "heavier load of bananas."

THE MAJORITY OF smugglers don't know they're smugglers. "Individuals bring back fruit for sentimental reasons," says Allen Clark of the Pest Exclusion Branch of the California Department of Food and Agriculture. "The profit motive is rarely involved and the activity takes place on a very small scale." At meetings held by the USDA to develop a two-way dialogue, new immigrants caught with fruits are educated on the risks of smuggling. The organizers have noted complaints that fruits don't taste good in the United States. Immigrants want to bring their tastier native fruits into America. Some of the attendees have asked whether it would help to wrap their fruits in cellophane before bringing them in. The USDA officials duly inform them that doing so still entails an act of smuggling that can land them tens of thousands of dollars in fines.

Some people will invariably continue to smuggle in a taste of home. There also exists a subset of conservationists who, fully aware of the rules they are breaking, smuggle for botanical reasons. For them, disseminating rare plant materials is a way of expanding a fruit's reach, of facilitating the dispersal of plants that might otherwise be faced with extinction.

Others, like Morimoto and his Japanese citrus clippings, smuggle for exclusive germplasm. The payoff can be quite lucrative: the 245 acerola seeds smuggled out of Puerto Rico in 1956 have blossomed into a major crop in Brazil. The seeds of a popular New Zealand citrus variety, the Lemonade fruit, are said to have entered the United States "informally." There are many ways to move contraband plants. One informal method is to falsify export permits. Voon Boon Hoe told me that many people simply mislabel the contents of their packages, saying they contain legal plants when they really contain all sorts of endangered jungle fruits. "The officials don't know the difference," he said. "If anybody asks you, you say, 'Oops—I thought that's what its name was.'" Another method is to use decoys. Smugglers will put a legit plant specimen in the shipping container, get the permit, and then switch the plants. Declaring a mundane fruit—apples, say—can also act as a distraction for something more controversial—such as a coco-de-mer. Perhaps the simplest recourse is to avoid declaring any seeds or cuttings.

That's how one California fruit grower brought home a rare variety of golden peach. He'd traveled to Tashkent to try to track down the fruits written about by historian Edward H. Schafer in *The Golden Peaches of Samarkand,* a history of exotica during the Tang dynasty. "The golden peaches actually existed," wrote Schafer. "Twice in the seventh century, the kingdom of Samarkand sent formal gifts of fancy yellow peaches to the Chinese court." But according to Schafer, these peaches had vanished without a trace. "What kind of fruit they may have been, and how they may have tasted, cannot now be guessed."

The grower told me he's certain that the golden peaches were actually nectarines. "How could something be golden *and* fuzzy?" he asked. "A peach can't shine. And the nectarines I found over there aren't red at all; they just have an amazing golden luster." Indeed, the small, shining fruits, now growing on his farm, were more golden than any nectarine or peach I'd ever seen. After finding the fruits in Tashkent, he'd pocketed a number of seeds and simply brought them home. "U.S. customs never asked about them. They didn't care. They wanted to know if I had Russian nesting dolls."

Growing certain fruits is often the only way to taste them: even if they were to somehow be grown on a large enough scale to merit phytosanitary research and access to irradiation facilities, many are simply too fragile to survive the indignities of shipping. The Chinese emperor Hsüan Tsung used to employ a special horse-riding courier to provide lychees for his Precious Consort Lady Yang. This fruit cowboy would race across the entire length of China, from Lingnan to a palace near Ch'Ang-An, bearing his royal consignment. Other fruits were shipped inside leaden containers filled with snow, such as watermelons exported from Khwārizm (an oasis in what is now southern Uzbekistan) and mare-nipple grapes transported across the desert from the Mountains of Heaven. But even by royal decree, much fruit cargo was so fragile that it couldn't be transported to the capital. Queen Victoria purportedly promised knighthood to anyone who could bring her a fresh mangosteen from Southeast Asia. No one, so the yarn goes, could fulfill the task. But the queen had other cravings she managed to sate. She was so fond of Newtown pippin apples from Virginia that she waived their import duties.

The idle rich have always had a penchant for unobtainable delights. Today, the most unusual subgroup of smuggler consists of the independently wealthy fruitleggers. These obsessive collectors are willing to pay whoever they have to and break whatever laws necessary to obtain the objects of their desire.

Consider the case of S, a portly, wisecracking plant aficionado in his early fifties. I first met him when Kurt Ossenfort and I were looking for a place to film fruit footage. David Karp suggested S's backyard, a jungle about the size of a football field in the upper enclaves of Bel-Air. Karp also told us that S lived with a pair of elderly sisters—spinsters who were the heiresses to some unspecified fortune. Mysteriously, they had adopted S as their legal heir.

Looking like Humpty Dumpty with his matted-down hair and tight XXL polo shirt, S waddled through the winding footpaths of his estate, pointing out cinnamon trees and palms covered with metallic spikes. "My bitch-tit tree is coming along nicely," he gloated, stroking a nipple-like appendage with his chubby hands. "Come, quats," he said, ushering us deeper into the forest.

He had just received a huge shipment of rare trees in jars, including one he delighted in calling "the testicle tree." He fixed me with a withering glare when I said he'd probably have trouble growing his mangosteen trees. "Oh, you're one of those people who thinks they know everything, aren't you?" he said, with the unflappable confidence of the overwealthy. William Whitman told me that nobody in America had ever brought a mangosteen to fruition besides him. I tried to ask if anyone had ever done so in California, but he'd already moved on. "Here's an African jelly fig I bought from an ugly old woman in Santa Monica," he said. "Oh look! My first finger-lime buds! Let's dance!

"These are some unknown palms that a woman smuggled in from China," he said, as we walked through the tangled greenery. "I forced her, at gunpoint, to sell them to me." I asked him more about the smuggling. "It's the only way to get some of the most remarkable trees," he continued. "We've brought many endangered, exotic species into this sanctuary." Caressing one particularly thorny trunk, he said, "I'm the only person in continental United States with one of these. We had it airlifted into the garden using a seventy-thousand-ton crane."

One of his strangest acquisitions was a fried-egg tree. "The shells of these fruits are used as protective penis covers by tribes in Africa," he explained, adding that his friends use it the same way at pool parties. He then invited me to witness this firsthand at an upcoming "bacchanalian revelry."

"Everyone who comes has to wear a penis shell," he said. "And we all take Chinese Viagra—'it's good for the male.'"

Although I never made it to one of those pool parties, I did accept his invitation, a few weeks later, to an afternoon barbecue in the jungle.

PARKING MY 1982 Acura from Rent-A-Wreck near the four bronze statues of Pan which stand sentinel at the entrance to S's property, I knock on the door. Nobody answers. I ring again. After waiting around three minutes, I push on the door, which is ajar, and enter the mansion. The living room is alarmingly cluttered. Stacks of dusty unopened envelopes, catalogs, postcards, stock appraisals, bank drafts and letters are heaped in corners. I pick one up at random; it's a legal notice from the early 1980s. Display cases full of trinkets overflow onto unhung paintings. The floor is a tangled heap of discarded scarves, rolled-up carpets and other debris. "Hello!" I yell. No response.

I head into the kitchen, where the circular table holds a gravity-defying mix of unwashed plates and cutlery, cracker boxes, chocolate containers, half-eaten bonbons and more junk mail. Cereal bowls, possibly from that morning, are precariously balanced on top of everything. The walls are covered with ornate landscape paintings and portraits of wan aristocrats.

Feeling like I've somehow landed in an alternate version of *Grey Gardens,* I head upstairs, calling S's name. No answer. I peer into a room, and one of the old ladies leaps out. "Don't look in here," she snarls. "Did you see something you weren't supposed to?"

"I don't even know what I wasn't supposed to see," I stammer.

Composing herself, she tells me that S is down at the picnic table. Going outside, I descend a spiral staircase surrounded by dozens of colorful bromeliads and reach a multitiered, man-made waterfall. Here's where things get bacchanalian, I imagine, looking at the lagoon under the waterfall. Nearby, an outsized hot tub is carved into the rocks. S later mentions that he is in the process of having another lake installed—for fishing—to be stocked exclusively with fruit-eating piranhas.

I descend into the heart of the forest, getting lost several times before finding S manning the grill in a picnic area breathtakingly situated inside a gorge. The walls of the nook are fashioned out of large pieces of quartz and other crystals. Shards of amphoras and ancient jugs are fitted into the facade. A refrigerator is built into the rock. A couple of ornamental giant clam shells are filled with crabs, lobster tails, oysters and other shellfish.

I sit down in the shade of Costa Rican palm trees. S describes the entire picnic area as "Costa Rica" because of all the imported flora. As I sip some fresh lemonade, S regales me with a few jokes.

"Did you hear about the guy with five penises?" he asks. "His pants fit like a glove."

The conversation turns to S's favorite local spots to get a "joe-blob." He recounts a recent trip to an adult cinema: "I was there and this guy

behind me says, '[S], is that you?' 'No, it isn't,' I said. It was this old German guy I used to buy antiques from."

He then starts talking about his all-time favorite names: Marina
Pickless, Arlen Snuckles, Bootsy Caucus. "There was Dick Tickle, a fur-
niture salesman," he said. "And Blithe Piddle, a decadent mink-and-
manure Virginian. Hussies and tarts, trolleys and tramps, trollops and
hustlers. I even knew a guy named Rich Tinky. He had fat, sweaty,
Pillsbury-doughboy hands and sold insurance. Who'd ever buy insurance
from someone named Rich Tinky?"

David Karp's name comes up. S calls him "the Freak Detective." "His
mom was so beautiful. I guess he has a recessive gene. In fruit terms, he'd
be called a mutant—no, a hybrid." He then tells a story about how Karp
had once invited him to a formal gathering. "On the way there, he asked
me not to embarrass him. I said 'Embarrass you how?' He said, 'You
always say offbeat things—just don't say anything offbeat.' So when I
arrived, the first thing I said to everybody was that 'Karp didn't want me
to say anything offbeat, which is fine because anything offbeat makes me
want to beat off.'"

A few minutes later, the sisters come hobbling through the ferns, and
the tone becomes more G-rated. They speak of S like a precocious child.
"He's so clever," fawns one of them. "He even writes screenplays and
Irish poetry that is very wise and spicy. You can only hear them at private
readings." S proceeds to recite one about a butterfly being a caterpillar
with wings that taste like powder.

As we eat, I ask how they met. "We used to bid on the same paintings at
auctions," says one of the sisters. "I'd always see them at flea markets,"
continues S. "They'd be driving around in a '62 Cressida and we were
always interested in the same things. We got to talking, and then one day
they invited me over."

S, the real fresh prince of Bel-Air, had made the ultimate thrift store
find. The trio started hanging out more and more, going on art buying
adventures together. Soon enough, S was coming over and taking care of
them—and their garden. In 1991, the sisters designated him as their sole
heir. Their lawyers spent an afternoon convincing them not to do it, saying that it was madness, explains S. "But the ladies knew what they
wanted. The lawyers said, 'Fine, but as soon as we sign the papers, he can
walk with everything.' The sisters said they were comfortable with that."
When the lawyers finally let S into the room, he acted very courteously—

until the papers were signed. "As soon as they signed, I stood up and yelled, 'You can get your own damn ride back home!' I left, slamming the door. And then a few seconds later, I peeked back in, just to see the look of surprise on everybody's face."

IN THE FOURTH century B.C., Ch'u Yuan wrote about why he loved fruits in a poem called "Li Sao," translated as "Getting into Trouble." As the popular saying goes, "a stolen apple always tastes better." The British have actual words for pinching fruits. "Scrumping," according to an online slang dictionary, means stealing apples from someone else's trees. The art of apple theft is "oggy raiding."

In certain circumstances, scrumping is entirely legal. The term usufruct refers to the right to use and enjoy something that belongs to another person when it extends beyond their property. The word comes from the Latin *usus,* to use, and *fructus,* fruit. It applies to fruit dangling from a tree into the street, or an alleyway or into another person's lawn. It's courteous to ask before plucking, but in case of a disagreement, usufruct provides a legal justification for eating them.

Thoreau was a firm advocate of scrumpers' rights. "What sort of country is that where the huckleberry fields are private property?" he howled. If only St. Augustine had known about usufruct, perhaps he'd have been less hard on himself—and the rest of Western Civilization. Book Two of his *Confessions* describes the night he and a band of fellow ruffians stole pears off a neighbor's trees. It was a thrilling sin—which later filled him with tremendous guilt. "Perhaps we ate some of them, but our real pleasure consisted in doing something that was forbidden." Jean-Jacques Rousseau's *Confessions* tells how he was beaten for stealing apples as a thirteen-year-old, an incident that marked him for life: "The horror of that moment returns—the pen drops from my hand." Lucky for him he didn't grow up in ancient Greece, where a law passed in A.D. 620 meted out the death penalty to fruit thieves and fruit tree molesters.

John McPhee, reporting on orange thefts in Florida, learned of burglars jumping out of anchored boats with burlap bags, picking fruits under the cover of night, and making getaways with thousands of stolen oranges in luxury sedans. One thief boasted that he could pick enough oranges by moonlight to fill a Cadillac in three hours.

A 2006 banana shortage in Australia caused by orchard-flattening cyclones led to a rash of fruit thefts. *The Times* of London reported that rustlers were breaking into unguarded plantations at nighttime and cutting off bunches of fruits. Bananas more than quadrupled in value. Fruit stores were also being targeted: one grocer put out a sign saying, "No bananas are kept on these premises overnight."

Safeguarding fruits has become increasingly vital. In Florida, sapote farmers guard their trees with rifles. Others pile fresh soil around their trees to track footprints. Bands of crop-heisting guerrillas roam rural Madagascar, leading farmers to stock firearms for self-defense. Corsica's kiwi Mafia are notorious for attempted murders of farmers unwilling to pay protection money. Avocado commandos have told Californian farm workers, "We're coming to steal these avocados, and if you don't like it we'll kill you." Three Flags Ranch, the biggest mango farm in California, occupies a sprawling 192 acres near the Salton Sea. The ranch's thirty thousand trees are entirely cordoned off with razor wire, a precaution taken after their first planting of mango trees was stolen. Not the fruits, mind you, but the actual young trees.

Stories circulate among fruit growers of industrial espionage, of spies breaking into seed banks and stealing rare clonal materials in the hopes of turning a profit. For that reason, the original navel orange tree and the Golden Delicious apple tree were enclosed within padlocked cages. Intellectual-property theft is rampant in the fruit world. According to the National Licensing Association, roughly one in three patented fruit trees are grown illegally. Several farmers told me that name "Dinosaur Egg" was stolen from the inventor of the Pluot when another grower heard him use the name, and quickly trademarked it. The recent wave of white-collar fruit crime has taken on an international dimension. "Stand By to Repel the Fruit Pirates" blared the headline of a Thai newspaper about western countries removing Southeast Asian fruit in order to breed improved varieties.

As I was writing this book, some disturbing crime occurred in my own Montreal neighborhood. A fruit store just down the street was fire-bombed on two different occasions, and no culprit was ever found. Some suggested that it was a mob hit, others that it was orchestrated by their main competitor—another fruit store a few doors away. After the second Molotov cocktail, the demolished store rebuilt itself and reopened again, but a cloud of distrust has descended upon the grocers.

There's certainly a violence and desperation to the produce industry. While in Shanghai's enormous wholesale fruit market, I was accosted by a young pineapple wholesaler from Hainan Province. "Fruit is a dangerous business," he said. He didn't believe I was a journalist, and was convinced that I was there as a fruit importer. He gave me a piece of cardboard with his contact info on it: "Pay attention to me!" he wrote at the top of the card. "Remember me," he yelled out as I was leaving. "Introduce me!"

Part 3

Commerce

Artocarpus incisa, breadfruit

9

Marketing: From Grapples to Gojis

This rendition comes to you by courtesy of Kaiser's Stoneless Peaches. Remember no other peach now marketed is perfect and completely stoneless. When you buy Kaiser's Stoneless Peach you are buying full weight of succulent peach flesh and nothing else.

—Evelyn Waugh, *The Loved One*

I N 1903, Isabel Fraser, the principal of an alternative girls' school in New Zealand, suffered a breakdown from overexertion. Leaving her woes behind, she set off into the heartland of China. Traveling up the Yangtze River, she came upon some trees bearing *yang taos*: dusty, fuzzy, ovoid fruits with glistening emerald interiors. Seduced by their flavor, Fraser set aside some seeds, determined to grow them at home. By the time she died in 1942, those few seeds had multiplied into hundreds of thousands of trees.

The Ichang gooseberry, as New Zealanders came to call it, became such a bountiful crop that growers started to export it in the years after World War II. Shipped overseas as the Chinese gooseberry, the fruit's binomial was greeted in America with McCarthyist sneers. No way was some pinko Chinese fruit going to make it in the land of apple pie. The fruit's other names—the monkey peach, the macaque peach, the vegetal mouse, the hairy pear and the unusual fruit—weren't much better.

Auckland packers realized that a catchier name was crucial to foreign sales. After settling on the Melonette, the possibility of American melon tariffs forced them to reconsider. At a brainstorm session, someone suggested the Maori word for New Zealand's national bird: the kiwi.

Soon enough, the kiwi began to soar; 1960s catalogs blared the news: "You had better order now; they are scarcer than screen doors on submarines." The kiwi's metamorphosis from obdurate, unsellable foreign weirdo to megafruit quickened the pulse of growers, shippers, sellers— and marketers. What other rubies were languishing in the dust, just waiting for a nickname to nudge them into mass production?

Orchestrating another smash like the kiwi wasn't as simple as investors assumed. For every mango and papaya there are countless voavangas and farkleberries that simply don't break through in Europe and North America. Lucuma, a yellow fruit from the Andes, much loved by the Incas, was predicted to be "the next major crop" by *Fruit Gardener* magazine in 1990. Nearly two decades later, it doesn't appear to have lived up to its hype. Still, Peruvian government websites offer investors the opportunity to sink funds into the industrialization of this "excellent/flagship" fruit. Putting lucuma on the map requires, according to one brochure, an initial investment of only $499,290.

Few fruits ever become permanent fixtures like kiwi, but new varieties keep trying. In the 1980s, large sums went into pushing the babaco, an Ecuadorian fruit related to the papaya. By 1989 it was called an "Exotic Failure" because people just weren't into it.

Another fruit that flopped was the *naranjilla,* also known as the *lulo.* A fuzzy golden orb with green flesh that grows on trees with enormous purple-veined leaves, is used to make a popular juice in Colombia, Peru and Ecuador. The Campbell's soup company invested years and millions of dollars attempting to popularize naranjilla in North America in the mid-1960s. Test-marketing of the juice garnered rave reviews, but the project was abandoned in 1972 because the juice's high price deterred consumers accustomed to cheaper domestic fruit drinks. In this era of boutique juices, naranjilla juice could make a comeback.

How do you make a hit fruit? There's no precise methodology. Financial success with new fruits requires such an intricate confluence of factors that it's as uncertain as raising a happy child. Servicing the cravings of a passionate few is a start. Isolating an immigrant demographic and catering to their taste buds also helps. Corporations term this phenomenon "nostalgic trade."

The football-shaped, red-fleshed mamey sapote grows in Cuba, as well as in southern Florida. Most Americans are unfamiliar with the fruit, but Miami's Hispanic population loves eating it fresh or as the main ingredient in a delicious milk shake called a batido. Most places outside Miami sell powdered mamey batidos, but fresh ones are light-years better. Still, because its shelf life is limited, fresh mamey seems likely to remain confined to Florida and the tropics.

It's often up to enterprising growers to create their own markets. Roger and Shirley Meyer grow three dozen varieties of jujubes, a mahogany datelike fruit much loved in Asia, on their Southern California farm. They decided to approach the produce manager of an Asian market and give him free samples. Soon enough they were selling briskly for $3.99 a pound. In fruiting season, they sell hundreds of pounds of jujubes a day.

Another shortcut to surefire profit is to legalize a previously outlawed fruit. Banned throughout most of the twentieth century in New York State, black currants weren't allowed to be sold, cultivated, transported or grown because of their link to a disease threatening pine trees. The interdiction was lifted in 2003 thanks to the lobbying efforts of a farmer named Greg Quinn, who now bottles a natural juice called Currant C. Whenever a banished fruit writhes out of its shackles, it stands to bring growers a significant "candy bar" (the agricultural term for a windfall). Studies forecast U.S. black-currant sales of 1 billion dollars annually.

In recent years, Californian farmers realized that Asian dragon fruits aren't allowed into the United States, so they started growing their own. They were all over New York's Chinatown in 2007. The simultaneous rise of travel to equatorial regions and the blossoming of foodie culture have led to increased interest in novel varieties. Historian Margaret Visser attributes this to a phenomenon called neophilia, the love of the new. Supermarkets, traditionally neophobic, are now starting to stock exotics because of the surging demand.

Just as Hollywood churns out sequels, kiwi growers are finding new ways to capitalize on the fruit's initial success. With golden kiwis making a splash in recent years, red- and purple-fleshed kiwis, as well as a variety covered with white polka dots, are now being introduced. Tiny hardy kiwis with edible skin, known as peewees or passion popper kiwi-berries, have also started generating solid revenues due in large part to their cotton candy flavor.

Proper promotion is vital, says Frieda Kaplan, a wholesaler who worked with growers to promote the original green kiwis. "For other specialties to develop like kiwi fruit has," Kaplan says, "growers must know how to manage their crops so that they can achieve a reasonable profit even at lower prices." She spent eighteen years publicizing kiwis by sending samplers to journalists, giving free taste tests, placing advertisements, working with growers and encouraging restaurants to incorporate them.

Chefs play an important role in promulgating new fruits. Martha Stewart has long been pushing white apricots. Meyer lemons are now well known thanks to their use by Chez Panisse's pastry chef Lindsey Shere. Alice Waters's cheerleading of mulberries has resulted in skyrocketing prices at California farmer's markets. When yuzu, a Japanese citrus, started being touted by celebrity chefs like Jean Georges Vongerichten and Eric Ripert, producers rushed to bring anything yuzu to supermarket shelves.

When chef-artist-scientist Ferran Adrià of El Bulli first tasted an Australian finger lime, the latest buzz fruit, he broke into tears. Each finger-shaped fruit teems with spherical pulp vesicles that are the citrus equivalent of caviar. When the skin is sliced open, these tiny translucent pearls glide out of their vacuum-sealed packaging. The peel's colors ranges from purple to crimson to alligator-green; the interior comes in shades of pink, yellow and nacre. Tasting them is akin to one's first sip of champagne; it's the sheer madness of the unexpected. As the finger lime's vegetal hype continues growing, commercial growers in California are setting up orchards. In a few years, they should be commercially available.

BEFORE LAUNCHING new varieties, marketers study consumer preferences. Multicolored pie charts reveal what percentages of shoppers like their fruit firm, soft, juicy, tangy, sweet, dry or moist. Hugeness, once thought to be a key goal, has proven undesirable. Bananas are morning fruits, strawberries are mainly evening fruits. Bananas, apples and grapes are fruits people like to eat on the go—others require preparation and are being prepackaged with that in mind.

According to surveys commissioned by the California Tree Fruit

Agreement, the most sought-after fruit demographic is a group called "Summer Enthusiasts." What unifies this sunny cabal, alongside their above-average fruit purchases, is an interest in playing sports and (say the following with a robot voice) having new experiences. Summer Enthusiasts "believe having fun is the point of life, think continuing to learn throughout life is very important, believe enjoying life and doing the things they want to do is important." Over 111 million Americans—in an estimated 53 percent of households—are Summer Enthusiasts.

Another important fruit buying subset is "Light Lifestylers"—people who are health conscious and like to exercise. Overlapping somewhat with the Summer Enthusiasts are 72 million "Super Moms and Dads"—the type who verify ingredients and nutrition stats prior to buying, and for whom family is everything. By far the most elusive segment of the population is "Generation Starbucks"—youngish people who still believe they are invincible (so health isn't a purchasing factor). These twenty- and thirtysomethings buy whenever the urge strikes them. Reaching the portion of this group with "positive life attributes" (ie., not the suicidal, bearded nihilists) requires making fruits available everywhere, like their namesake java.

For all of these various menageries, fruits are being pushed as a break-time snack. Branding gurus want to make fruits a part of transitions from one activity to the next: rejuvenating tide-me-over breaks, mid-afternoon pick-me-ups and after-work snacks. To merchandising reps, it doesn't really matter when these moments happen as long as they become ritualized routines filled with fruit. Once fruits have become ingrained as the go-to transitionary fuel, their multisensory experiential qualities can be leveraged into high-volume snacking. Or something to that effect.

Fruit catchphrases and slogans bandied about in these studies include "little taste adventure," "delicious handful of goodness" and my favorite: "the snack that quenches." Like a lingering stereotype, these studies' undertones are shrouded in a gauzy cloak of believability. After poring over them for several hours, I badly need a transitionary moment. Despite, or perhaps because of, all the viral jargon, I feel an urge to go to my fridge and eat a peach. It certainly appears to be a "guilt-free treat," rather than an endorsement of big agriculture. Then I bite into it, my teeth sinking into what feels like wet sand. Not quite the Burst of Fun I was hoping for.

The power of branding hasn't been lost on growers. The "Delicious" apple, a precursor to Red Delicious and Golden Delicious apples, started as a name in a marketing boardroom. Only subsequently was an apple found that fit the bill. The tactic worked: the Red Delicious was the most successful apple of the twentieth century. But its era is over. According to organoleptic experts, as the apple became redder, it also became blander. In the midst of its downfall, one American nursery has used marketing to concoct an entirely new breed of fruit.

IT'S AUTUMN and ripe apples are being picked in Wenatchee, Washington, the Apple Capital of the World. From a distance, the red orbs glowing in their boughs resemble Christmas tree ornaments. But as I pull into Gary Snyder's orchard, this wholesome image starts to warp. Something smells like grape bubble gum.

"Sorry—it's me—I've been doing Grapple testing," explains Snyder, the inventor of an apple that tastes like grapes. "The percentage of solution I get on myself can be quite high when I work with the raw product." Snyder has never revealed how Grapples are made, saying only that it involves dunking Gala or Fuji apples in artificial grape flavor. This chemical solution is so potent, he allows, that one tainted T-shirt in the laundry can spread the aroma to all the other clothes. Wherever Snyder goes, a saccharine cloud follows. "It's getting to the point where it's getting tiring for the wife," he says, the sunlight tinting his glasses an ashy shade of merlot.

Snyder, forty-five, is the marketing director of C&O Nursery, a company owned by his family that sells tree stock to farmers. He has a pudgy face, squinting eyes and a sloping forehead topped with bristly brown hair. He's wearing sneakers, little ankle socks, khaki shorts and a polo shirt embroidered with the Grapple logo. He wears a gold ring on each hand. With those tinted glasses, he looks like Dr. Strangelove on a casual Friday.

When I ask about his earliest memories of eating apples and grapes, Snyder takes a long pause. When he finally speaks, he says the first apple he ever tasted was a green, underripe Granny Smith in his family's test plot. And opening up the grape door, he starts telling me about childhood summers when he and his brothers would get lost in their father's Concord grape patch: "I remember eating them until I was sick."

I pluck a Gala and wipe off its powdery white residue. It tastes fine, despite that cloying odor. I can't tell if the smell is wafting off Snyder, or the apple or if it's coming from somewhere else. He's trying to assure me that it's only coming from him, but when he drives off, a whiff of deception trails in his wake. Even after he and his perfumed clothes are gone, the orchard still smells like synthetic grapes.

IN AN EDITORIAL about the fruit, the *Ann Arbor Paper* pondered the point of eating something whose name means "to come to terms with." In actuality, Snyder's fabrication is pronounced *gray-pull,* not *gr-apple. Jeopardy* contestants were stumped when the fruit was featured as a Daily Double question in 2004: "From Washington, the Fuji type of one fruit flavored with the Concord type of another yields this, also meaning 'wrestle.'"

Complicating matters, there is another fruiting plant called the grapple (the pronunciation is, in this case, homonymous). Also known as devil's claw, the grapple (*Harpagophytum procumbens*) has evolved the peculiarity of clamping itself to the feet of ostriches, which ensures the dispersal of its seeds. I made the mistake of rhyming Snyder's creation with "apple" the first time we spoke, and was swiftly upbraided. Later, in what I suspect to have been a retaliation, he called me "Alan"—emphasizing it, pausing and slowly turning to observe my reaction.

Grapples are sold with the tagline, "Looks like an apple. Tastes like a grape." A four-pack retails for around four or five dollars at Wal-Mart, Safeway, Albertson's and other chains. More than 36 million Grapples have already been sold. Clearly, a lot of people like them. Most people react with a " 'Wow,' " says Snyder. "Your mouth is telling your brain something's wrong."

Despite its platinum sales, the fruit hasn't entered the marketplace without resistance. A kiddie taste test organized by the *Denver Post* deemed it "ucky." Eating one calls to mind biting into lip gloss. It is not organic, nor is it certified kosher. On a website where users can leave their feedback, the comments are often vitriolic. "I am disgusted," writes one. "The fact that these people have the audacity to charge $5.00 for Dimetapp-flavored apples is just wrong. The inventor should be shot in the foot." The litany of complaints continues: "Yuck! I can't believe these folks are marketing this crud. . . . Blech!" Someone named Trevor writes: "I ate a grapple last night and i was kinda scared about it. I mean who knows what's in them because it sure doesn't say in the ingredients. After i bit it i was confused . . . just go buy some Fuji apples and it tastes the same and you save money and you don't run the risk of dying from whatever crazy chemicals are in them."

The only ingredients listed on the packaging are apples and artificial flavor. Until 2005, "fatty acids" were also included in the labeling. "I wouldn't have known that there was fatty acid in the Grapple except for the fact that I have a very bad reaction to fatty acids—explosive diarrhea," writes "Pete" online. "I was on a road trip with my family one minute in the middle of a snowstorm and the next minute I was in the car in the middle of a shit blizzard."

Snyder dismisses these protestations as mere sour grapes: "If five percent of people complain, I've got no problem with that. People who don't like it probably wake up with a cranky attitude. Fine. You're not going to make everybody happy. If you don't wanna eat artificial flavoring, what do you wanna eat? Everything has artificial flavoring today, except a banana or something like that."

The Grapple's critics counter that apples, like bananas and other natural things like that, are not improved by the addition of chemical flavors. Fruits, they argue, aren't meant to be industrialized. But as our foods have become standardized commodities, producers like Snyder are merely giving retailers what they demand: homogeneity. Calibrated for even size, and processed when they attain a sweetness level of fourteen Brix, all Grapples taste identical, whether eaten in Wenatchee or Witchita. "It's not just a burger, it's a McDonald's burger," says Snyder. "It's not just an apple, it's a Grapple."

Snyder, who invented the Grapple in 2002, believes that creating it has been his divine calling. "Someone had to bring this out," he says. "I was the one who was chosen to do so." Although he becomes guarded when asked how he got the idea for the fruit, he claims that the idea for a flavored apple evolved over time. "There was no lightbulb moment," he says, waving his hands in the air, patting his belly and drawing circles with his fingers. "There were multiple things involved that I can't get into here for trade reasons. There was no epiphany."

HISTORICALLY, Washington fruit production can be traced back to a eureka moment. The first apples and grape seeds planted in the area arrived in the vest pocket of Captain Aemilius Simpson in 1826. In the service of the Hudson's Bay Company, the young captain attended a farewell banquet in London before undertaking his lengthy voyage to the wilds of the Pacific Northwest. The dessert course consisted of fruits. Conversation turned to foods of the New World, and to the fact that no apples or grapes were then growing on the Western frontier.

At that time, Johnny Appleseed was still spreading apple seeds along the Ohio River. Grapes, such as the Scuppernong, were native to the Eastern seaboard, but weren't yet cultivated in Washington. The range of the Frost grape, another widely dispersed indigenous variety, extended no further west than Montana. To give an idea of the scarcity of fresh fruit on the Pacific coast at that time, apples from Oregon were sold in San Francisco in 1850 for $5.00 apiece. That year, the average monthly wage for an unskilled laborer in California was $4.20. Calculating based on the nominal GDP per capita, $5.00 from 1850 equals $1,911.75 current dollars—not far off from the average monthly wage of an entry-level worker today. (According to a 2001 study, half of all American farm workers earn less than $625 per month.) Granted, 1850 was the height of the Gold Rush, and the wealth twinkling out of the earth permitted astronomical expenditures on luxury items such as apples.

Back in London, a young woman sitting near Captain Simpson was overcome by the symbolism of his imminent voyage to a barren land. In a romantic gesture, she slipped the seeds out of her apples and grapes and placed them in Simpson's pocket so that he would be able to plant them upon his arrival.

Several months after landing in Fort Vancouver, on the banks of the Columbia River in Washington, Captain Simpson was the guest at another dinner party. Wearing the same suit, he remembered the pips, and presented them to the fort's leader. The "love-seeds" were planted in the spring of 1827. This intertwining of apples and grapes marked the beginnings of Washington's remarkably fertile fruit production—an American agricultural success story that has, in recent years, taken an unexpected turn.

YIELDING 12 BILLION fruits each fall, millions upon millions of apple trees sprout from the desert east of the Cascade Mountains in central Washington. Throughout most of the twentieth century, Wenatchee truly was, as its town motto has it, the apple capital of the world. Their Red Delicious beauties were sold around the globe. The United States produced more apples than any other country, with Washington dominating domestic levels. Then, in the early 1990s, Chinese orchards erupted. China now produces 25 million metric tons of apples per year, compared to the United States's 4.3 metric tons. Other countries have also jumped on the apple wagon, their diminished labor costs cutting into American sales.

The U.S. market collapsed in 1997. In the face of reduced exports, cheaper foreign imports, stringent inspectors and tariff wars, numerous Washington apple growers have declared bankruptcy or had their orchards seized in foreclosures. As farmland gets snapped up, consolidation has led to fewer, larger agricorporations controlling the industry, leaving little room for independent farmers. Eking out profits from the constricted pricing margins requires not only high production volume, but a stake in other aspects of distribution. Apple growing today entails vertical integration, with companies simultaneously involved in growing, sorting, packing, storing and shipping.

The Grapple was formed in this crucible. With farmers either giving up or branching into different aspects of production, peripheral industries have had to adapt. Established in 1905 by Snyder's grandfather's uncle, C&O Nursery is now the oldest active nursery in Washington, and one of the oldest in North America. Surviving in business for a century requires evolving with the times. As everyone in Washington knows, the only hope for the state's ailing industry is to find ways of getting people to eat more apples. Annual American per capita apple consumption is 15.1 pounds, about a third of the European average (around 45 pounds), and well under world-leader Turkey's 70.77 pounds. The Snyder family hopes that their flavoring innovation will help boost sales of regular apples as well. They call this process a "chain-reaction of apples."

They're also tapping into another new trend in fruit marketing: non-hybrid hybrids. The Grapple comes on the heels of the Strawmato, a very sweet tomato masquerading as a cross between a strawberry and a tomato. Strawmatoes have no strawberry genes; they are united in name alone, although whether you say *straw-mate-o* or *stra-motto,* it just sounds off.

Mango Nectarines, according to their website, contain "a hint of tropical flavor that's reflected in their name." Ito Packing Co.'s oblique insinuation that their nectarines actually taste like mangoes has angered some fruit professionals. "They're capitalizing on naivete," fumes heirloom stone fruit grower Andy Mariani. "It tastes nothing like a mango, not that a cross between a nectarine and a mango is even possible. It's just an anomaly from their genetic scrap heap that happens to be entirely blush-free, so it sorta resembles a yellow mango."

Though Mango Nectarines taste like regular nectarines, their novelty commands a higher price. Their flavor is better than another Ito Packing Co.'s innovation, the Honeydew Nectarine. Its green-tinged white skin certainly bears a resemblance to the exterior of a melon, which raises logistical concerns for the neophyte: namely, will the rind be edible? Of course it is, because it's simply a white-skinned nectarine with a trumped-up nom de plume.

Such titular hybridization is a double-edged seed, as it both attracts new customers, but also leads to alienation when the duplicity is revealed. Once shoppers catch on, the possibility of a backlash could relegate the fruit back to obscurity. Economists have noted the consumer tendency to be "punitive" when they buy a disappointing fruit: they won't buy any more of that fruit for a while, and in some cases they give up on that variety forever. "If you eat a bad cherry, it can take you six weeks to go back and buy another one," says Snyder. "By then the season's over. Too bad, try to get them next year." In the short term, however, flashy *fruits-du-jour* like Strawmatoes and Mango Nectarines—or Grapples—can be counted on to generate profits.

"I know they're unlocking the DNA of the apple right now," Snyder says, leaning back in his chair. "Everybody wants to use gene machines. If you take a sweet gene and a red gene and you put 'em together—you got a sweet red something. Does that scare you? I had a cauliflower-cheese soup the other day. I was just talking about it at the store, about how it's like a Grapple. I don't like cauliflower by itself, but once you add cheese . . . There are a lot of things that really go well together. Cousin Todd says it best: 'chocolate and peanut butter.'"

C&O Nursery subscribes to a number of trade magazines, both of the fruit-growing variety and of the chemical-flavoring variety: *Food Technology, Produce Business, Good Fruit Grower, Food Chemical News.* Their diverging editorials seem to have intersected in Snyder's mind. "You wanna have a Sudafed apple in the future? I'm not gonna close any doors," he says. "Look at nanotechnology. We don't know where it's gonna go."

In recent years, the advent of presliced apples has been a boon to growers. Studies have shown that 65 percent of consumers would rather purchase sliced apples than whole apples. Plastic baggies now sell in enormous quantities at grocery chains and fast-food outlets like McDonald's (which purchased 54 million pounds of Galas in 2005 to make Apple Dippers). Produced in factories around the nation, the fruits are mechanically cored, sliced with scalpel-sharp steel blades, sorted on assembly lines by workers wearing head-to-toe protective outfits and dredged in an invisible, flavorless powder made of ascorbic acid, calcium salts and vitamin C called NatureSeal. (NatureSeal has also become available for domestic use, meaning apples can now be cored, sliced and doused in the comfort of our homes. "It works wonders," says a sales agent.)

Conforming to the sterility requirements of processed food laws isn't simple. The packages can harbor listeria, salmonella and toxic microflora causing diarrhea, mucosal invasion and carcinogenesis. Outbreaks of bacterial contamination have already forced sliced-apple recalls. New chemical additives such as PQSL 2.0 are being developed to neutralize all traces of life on the apple slices.

Grapples have also been available presliced in baggies, but Snyder says they may not be for long. C&O Nursery recently stopped working with their initial marketing partner, Get Fit Foods (who own proprietary slicing technology). Before handing me a press release, Snyder blacks their name out, first with a pen, then with a Sharpie, saying, "That's how much I like them."

"We came across divergent interests," is how Get Fit Foods coowner Blair McHaney explains their separation, noting that there is a "consumer groundswell" against the Grapple's use of synthetic flavoring on a natural product.

Get Fit Foods recently filed for patents on presliced apples containing natural flavors. "I think the mind is more accepting when it's *natural* flavoring," says McHaney. Apple Sweets, as McHaney's product is called, have been tested with forty different so-called natural flavors, such as caramel, root beer and wild berry, and are already being launched in supermarkets.

Posterity will determine whether flavored fruits are here to stay. For now, as the Grapple's sales testify, shoppers have cottoned to the novelty factor. Point three in a bullet-point document Gary Snyder printed up to prepare for our interview claims that a high percentage of people say it is the best apple they have ever eaten. "It's the only apple that gets fan mail," he says. "We get tens of thousands of e-mails. We're trying to keep it cool, to not toot our own horn—but we're going global. It's a smoker!"

IN THE REALM of flavored fruits, the Grapple's only analogue seems to be the maraschino cherry. "I like to tell people a maraschino is the nutritional equivalent of a LifeSaver," says Josh Reynolds, vice president of Gray & Company, the world's biggest maraschino cherry manufacturer.

Once upon a time, a marasca was a bitter cherry that grew wild in Croatia's Dalmatian mountains. The kernels were crushed and fermented to make maraschino liqueur, in which whole marasca cherries were preserved. It became a fad to have maraschino cherries with cocktails in New York in the early 1900s. Their quality has steadily decreased with each new manufacturing innovation. Maraschinos today are more like chemical-addled receptacles than actual fruits. Producing them involves bleaching low-grade cherries, stoning, sweetening and flavoring them, packing them in a syrup, and dying them—usually neon red, although Eola Cherry Co. also sells shocking blue, green and pink maraschinos.

The dye used until the 1990s, Red Dye #3, is now banned by the FDA. Research linked the color to cancerous tumors in rats. These days, maraschinos contain Red Dye #40, a pigment found in Doritos, Pop-Tarts and other foods. (Parents of hyperactive children claim that the dye causes tantrums.)

Snyder wrinkles his nose when I bring up maraschinos: "That's a bleached-out way to dump bad cherries." But when I ask him about the Grapple's safety, he reaches into his pocket. "A cell phone is a good example. What's a cell phone going to do to me? We don't know." He holds his phone to his ear questioningly, as though he might hear an answer or, perhaps, the ocean.

Part of what makes Snyder's contrivance so compelling is the secrecy with which he discusses it. His answers are peppered with talk of pending patents, levels of confidentiality and intellectual property issues. "We're in experimental lockdown," he says. "I have to tiptoe around what I'm saying and hem and haw about what information is released. My partners and I have signed over one hundred fifty NDAs [nondisclosure agreements] on this."

When I first contacted Snyder, he said, "No way on God's green earth will I let you into the plant." Only after a series of phone calls, and my offer to sign an NDA (which never materialized), did Snyder finally consent to an interview. I asked him about the process of making Grapples. "There's no reason to go down that road of showing how it's made," he said, his evasiveness igniting my curiosity. "It's far better to just do a happy story on how this is a way to keep kids healthy and to help with childhood obesity. Every time we tell a journalist that that's the story they should do, they're very happy doing that story."

Snyder said he would give me the *"Reader's Digest* version" when I arrived. "It's like asking how chocolate chip cookies are made," he said. "You blend 'em up, throw 'em in the oven and bingo, out comes a cookie. That's how Grapple works. We aren't going to give out the recipe."

One of Snyder's peculiarities is a tendency to make exaggerated gulps, or the sound of a gasket blowing off steam, especially when faced with a difficult question. During our first phone conversation, he told me "point blank" that he highly doubted I would learn anything by visiting him in Wenatchee, punctuating the statement with a glottal thud. "But if you do get a blade of grass, then at least it'll be something."

SET BACK FROM North Wenatchee Avenue, C&O Nursery's office facade consists of a series of one-way mirrors. They can see out, but you can't see in. Non-transparency seems to be an architectural constant in Wenatchee: windowless gray fruit packing and storing facilities dot the area.

The first thing I notice on Snyder's desk is a half dozen peaches (labeled Vergil and C-1XO) in a temperature-controlled container. Starting the conversation, I casually ask about them as we sit down in his office.

"Oh, you can't look at those," he says, covering them with his hands. "Those are new varieties."

This first exchange sets the tone. Later on, I ask if their lab is located in the building.

He stares at me. "Maybe."

"Can we go into the lab?"

"Nah."

Early in the interview, Snyder's cousin Todd Snyder comes into the room to listen in on our discussion. Shortly thereafter, Gary Snyder lets it slip that, "The trick is getting the lenticels to open up to get the flavor into the apples." When I inquire further about lenticels, the cousins exchange nervous glances.

Lenticels are the little specks on the surface of apples and pears. They are actually pores through which pome fruits breathe. Thoreau, in his manuscript *Wild Fruits,* described seeing heaven in the dots on the surface of a pear: "(As on the sky by night) the whole firmament with its stars shines forth." The Snyder family saw something else in those tiny openings: a big opportunity.

The blogosphere froths with attempts at explaining how Grapples are made. Some claim to have "noticed that the skin was perforated with multiple, almost invisible injection needle holes." Snyder dismisses this notion, saying that puncturing apples would immediately render them defective. One chat room explanation of the process alleges that the fruit is first immersed in a gelatinous substance, filled with a bacterial grape-flavor gene, and then rinsed with a concentrated liquid derived from Prozac.

When I read this to Snyder, he starts tapping his foot impatiently. Upon hearing the part about the antidepressants, he laughs uproariously. "That's why it's the happy apple!" he bellows, his face reddening.

"I never wanted to let anyone know anything about this," he says. "I like leaving it open. We used to get orders for Grapple *trees*. It's so much better when you let your brain go."

When I ask him to explain the process, he says his legal counsel has advised him to only discuss what's on the Grapple's website. It contains scant information. Several times over the course of our interview, Snyder responds to my questions by holding up his finger. Swiveling in his chair, he moves over to his computer, whistling a little tune. He then prints out a webpage showing an image of a smiling apple diving into purple water and hands it to me with an exaggerated smile. "That's the information that the public can have access to."

DOING SOME OF my own research into artificial grape flavor, I learn that the scent of Concord grapes has a chemical name: methyl anthranilate (MA). Naturally occurring in Concord grapes, MA is the substance that makes them smell the way they do. Every time we eat those grapes, organic MA molecules invade our olfactory glands.

MA can also be synthetically produced, resulting in artificial grape smell and flavor. Chemically constructed MA molecules are called "nature identicals." Flavorist lore holds that a German scientist discovered the compound $C_8H_9NO_2$ accidentally: when combining different chemicals, he was struck by the fragrance of grapes wafting from his Bunsen burners.

MA extracted directly from grapes (in other words, *natural* grape flavor) is more expensive than synthetic MA obtained through industrial pathways. PMC Specialties Group is the only producer of synthetic grape flavor in the United States. At PMC's chemical plant in Cincinnati, methyl anthranilate molecules are vacuum-distilled in a four-story refinery whose gleaming chrome pipes recall an El Lissitzky painting.

Used to enliven everything from candy to perfume, MA is in all products containing artificial grape flavor. MA is in purple Kool-Aid, grape soda, bubble gum and many other foods. Categorized by the FDA as GRAS (Generally Regarded as Safe), it has been used for decades without any evident signs of toxicity in humans—besides causing serious, albeit temporary, damage if it gets in our eyes.

While googling MA, I come across something as seemingly random as that German scientist's discovery: a website for Bird Shield Repellent Corporation, a Washington outfit that markets a pesticidal formulation containing MA as its active ingredient. Grape-flavored pesticide? It doesn't seem to make any sense.

I call the owner, Fred Dunham, who tells me that birds loathe the smell of MA. "Our storage warehouse used to be so full of birds that they'd give people umbrellas so they wouldn't get bombed from above," says Dunham. "When we started storing methyl anthranilate in the warehouse, the birds disappeared."

I ask him if his product is a pesticide. "Anything that kills or repels is by definition a pesticide," says Dunham. "So Bird Shield officially is a pesticide, even though it is nonlethal." The same MA used as an ingredient in our foods is also used in this food-grade bird repellent. Bird Shield's scent only lasts about a week, so numerous aerial sprayings are recommended throughout the growing season. It is used on a variety of crops, such as corn, sunflowers and rice. It is also used on certain apple crops.

Such as?

"Mainly Fujis and Galas."

Those are the two varieties used to make Grapples.

Has Dunham heard of the Grapple? "Sure," he replies. "They buy their methyl anthranilate from us. We make them their own formulation, which is, of course, confidential."

C&O NURSERY has more than twenty fruit patents, starting with 1932's Plant Patent #51 for the fuzzless Candoka Peach up to 2001's PP #12098 for their striped Top Export Fuji Apples. The family's latest patent, for "grape-flavored pome fruits," is still pending, as anyone trolling the databases of the United States Patent & Trademark Office can see. Everything relating to the Grapple's manufacture is contained in the various dockets of Patent Application #20050058758, filed by Gary Snyder.

The documents describe how MA is absorbed through the lenticels. It turns out that MA acts like a solvent, seeping into the fruit's interior when dipped in a 70 degree admixture. Large-scale treatment can incorporate a spray system mounted to a conveyor belt, but this purple rain method isn't as effective as fully dunking the fruit in packing-line dip tanks. After drying, the flavored fruits are placed in refrigerated chambers where they retain the flavor for months. The smell is volatile and can dissipate rapidly if the apples aren't stored in dark, cold places. This would explain the many online complaints that Grapples aren't even redolent of grapes—they hadn't been refrigerated. (The grape-scented papers Snyder gave me stopped giving off fumes after a week.)

The flavoring admixture used in the patent application is an off-the-shelf solution of MA "marketed under the names of Bird Shield Bird Repellent, and Fruit Shield Repellent." (Dunham's company also manufactures Fruit Shield Repellent, which likewise contains MA as its sole active ingredient.)

Item 0013 spells it out unequivocally: "For the present invention, the heretofore considered repellent effects of the methyl anthranilate are preserved for consumer enjoyment."

Yet when I call Snyder to clarify the use of artificial grape flavor in food-grade pesticides like Bird Shield, I get nowhere. "I know the road you're going down," he says. "See your word, 'pesticide.' I will not associate it down that road. It's not associated, period. It's a stop sign for me. I will not let it go to press." I can see why he opposes the word *pesticide*; nevertheless, MA is a flavoring agent that isn't harmful to humans. It just happens to repel birds, which makes it a pesticide.

As secretive as he may be, it's hard to dispute Snyder's contention that a Grapple is better than a deep-fried Twinkie. "I really think it will do good for mankind," he says. "It can help people. If you make a buck it's a bonus. If you don't make a buck you can't do it." The freakiness of his invention, while reminiscent of Dr. Frankenstein, owes less to the original than to the ethos of Mel Brooks's *Young Frankenstein* (wherein Gene Wilder's character insists "It's pronounced *Fronken-schteen*").

"This has brought more smiles to people's faces than everything I've ever done," beams Snyder. "Beats the heck out of a comedy club." When asked if he has ever worked at a comedy club, he says he hasn't—although correspondences from C&O Nursery are likely to sign off with "Have a grape day."

Other flavors are in the pipeline. Snyder, who claims to have "lock-jaw" on the subject, nonetheless says we can expect some "*berry* delicious apples" soon. "You can read whatever you want into that," he adds, with an exaggerated wink and a slap on the knee. "When you're in a supermarket, it smells gross when you walk into the chemicals aisle. It smells gross when you walk into the tire aisle. But when you walk over to Grapples," he says, flashing a thumbs-up and rippling a zephyr of sickly sweet grape odor, "it's killer."

"THE FUTURE OF my hometown," narrates Guy Evans in *Broken Limbs,* a documentary about Wenatchee's apple farmers, "is very much in question." As the city's orchards are being turned into subdivisions, big-box stores or land for other crops, local agriculture commissions are starting to promote the fact that they now yield more potatoes per acre than Idaho. (Even though Idaho still produces the most potatoes.) *The Wenatchee World* newspaper's slogan is equally convoluted: "Published in the Apple Capital of the World and the Buckle of the Powerbelt of the Great Northwest."

North of C&O's office, orchards hug the Columbia River all the way to the Canadian border. Fewer and fewer belong to small growers. One of the farmers interviewed in *Broken Limbs,* Dave Crosby, says that he first got into apple farming in the 1970s because he had been told that, "You don't have to work—you just pick the apples. You make lots of money and you can just have a good time." It didn't work out quite how he imagined. Crosby lost his orchard in 2003.

One of the main thoroughfares in Wenatchee, lined with the same fast-food franchises found all over America, is called Easy Street. It's a name dear to apple growers' hearts, especially to those who have adopted a monoculture approach. But no matter how many chemicals we throw at our food, humanity can't weed-whack away the curse placed on Adam after eating the apple: "In the sweat of thy face shalt thou eat bread, till thou return unto the ground." Bleeding rainbows of iridescent oil, the only Easy Street around here is littered with used burger wrappers.

And just as a crisis in apple growing led to Grapples, crashes in other fruit prices are leading to similarly intriguing solutions. "Growing cranberries used to be a pleasant way of life," says Hal Brown, moderator of a cranberry discussion group at www.cranberrystressline.com. "It's pretty, it's scenic, you could make some money, hire an employee or two, pay them a 30k salary. Unfortunately, the fruit's value declined from eighty dollars a barrel in the 1990s to twelve dollars a barrel in 2001. If prices had stayed up, we could have easily made 300k per year. Thank God my wife is a librarian and I'm a psychotherapist."

Brown has led the charge against a misguided attempt by Ocean Spray to market white cranberries as a new variety of cranberry. The original label on their juice stated: "These all natural, *fully ripened,* white cranberries come from the first harvest of the season so they're milder than traditional red cranberries." Brown wasn't impressed. "White cranberries are just unripe berries," he says. After he complained about these "outright lies" at the Federal Trade Commission, the language on the label was altered. "There's no dirty secrets about growing cranberries," he says. "It's the marketing of them that has the dirty secrets."

Ocean Spray has embarked on major advertising campaigns, coming up with an interactive concept called "Bogs across America." Every fall, they construct bogs in major cities to demonstrate how cranberries are harvested under several feet of water. The idyllic swamps full of farmers in thigh-high rubber boots contrast with the industrial sorting methods cranberries undergo once they're knocked off their low-hanging vines by tractorlike machines called egg-beaters. After being vacuumed into trucks and sorted on Bailey separators, they end up on a conveyor belt that sends them cascading over a precipice, sort of like a cranberry waterfall, where they're filmed by optical sorters equipped with over a hundred cameras. At the slightest hint of an imperfection, a precision gun shoots out a burst of air that blows any damaged berries out of the lineup. The next level of verification takes the berries into a room lit with ultraviolet lights—bad berries emit a fluorescent glow that allows them to be picked off the conveyor belts.

Much of Ocean Spray's marketing aims to recapture market share that they've been losing to pomegranate juice. Pom Wonderful juices are the brainchild of Stewart and Lynda Resnick, married Beverley Hills billionaires. They planted over a million pomegranate trees to capitalize on the fruit's healthy image. The Resnick's pomegranates are being sold with the promise that they "might even save a life. Yours!" Their slogan is: "Cheat death."

Some of the other claims surrounding pomegranates are as faith-based as Mohammad telling his followers to eat pomegranates because "it purges the system of envy." Pomegranates do contain myriad beneficial minutiae: tannins, antioxidant polyphenols, ellagic acid and punicalagin. But as the FDA-enforced warning states on one pomegranate-capsule maker's website: "These products are not intended to diagnose, treat, cure or prevent any disease." That hasn't stopped new products like condoms soaked in pomegranate juice from being sold as having extra protection against HIV. Experiments on rabbits have shown that pomegranates positively affect erectile dysfunction—a condition not normally associated with bunnies.

Whenever news reports come out saying a fruit or juice is amazingly healthy, the corresponding study is almost invariably funded by producers with vested interests. Researchers from Boston's Children's Hospital examined the results of 111 juice and beverage studies financed by manufacturers, and—big surprise—they found that the studies were often biased. If you read the fine print on the *American Association for Cancer Research Journal*'s article on pomegranate juice's effect on prostate cancer (which concludes that further study is necessary), you'll see that the study was financed by a grant from the Resnicks.

Nutrition is a shifting field riddled with contradictions, misbeliefs and fantasies. Despite the unanimity among nutritionists that we need to eat a variety of fruits and vegetables daily, as well as exercise, most of us don't. Entrepreneurs, capitalizing on misinformation, are using the power of marketing to magnify certain fruits' curative effects. As a result, vast amounts of money are being spent in the hope that individual fruits can heal all woes.

A 2006 ADVERTISEMENT in the *Farmer's Almanac* for a book called *Unleash the Inner Healing Power of Foods* claims that two servings a week of grape juice fights "heart attack, stroke, diabetes and cancer!" Its main source is a 1928 book called *The Grape Cure* that has sold more than one million copies and is still in print today. Written by Johanna Brandt, the book recommends that sick people should eat nothing but grapes. The author provided anecdotal evidence of curing terminal cancer patients. She told of a woman in the Bronx who couldn't stop vomiting day and night until Brandt administered some grapes. Another young woman with rectal and colon problems started the Grape Cure and began oozing pus. "When she began to pass worms, I knew that the terrible ordeal was nearly over. The grapes seemed to ferret out the most deep-seated cause of trouble and drive it from the system . . . Sanctified by suffering, this woman has emerged from the abyss of premature death to be a witness to the divine healing properties of the grape."

The American Cancer Society investigated the Grape Cure on a number of occasions and never established any proof of it healing cancer or other diseases. Quackwatch.org concludes that it is a book worth ignoring: "There is no scientific evidence that the Johanna Brandt's *Grape Cure* has any value."

Following grapes, there was an unsubstantiated craze for apricot seeds in the 1960s and 1970s. This mania can be traced to reports of the Hunzas, a remote people living in the Himalayas who were renowned for their longevity, endurance and lack of diseases. Apricots were the Hunza staple, used fresh, dried, crushed, baked and in an oil extracted from the kernel. As a result, a chemical compound in the apricot kernels, called Laetrile, came to be marketed as nature's anticarcinogen. Celebrity advocates such as a dying Steve McQueen as well as other desperate cancer patients latched on to Laetrile as a final hope, but it never seemed to cure anybody. Eventually, the National Cancer Institute sponsored a study to determine Laetrile's effectiveness. They found nothing. *The New England Journal of Medicine* declared that "the evidence, beyond reasonable doubt, is that it doesn't benefit patients with advanced cancer, and there is no reason to believe that it would be any more effective in the earlier stages of the disease." As a result, the sale of Laetrile was banned, and some of its marketers incarcerated. It continues to be sold online from clinics in Tijuana. In recent years, the United States has cracked down on companies such as Holistic Alternatives and World Without Cancer, Inc., that have sold Laetrile.

The latest fruit panacea to hit the market has also been linked to the Hunza. According to a booklet called *Goji: The Himalayan Health Secret,* tests involving infrared molecular bonds, a spectroscopic fingerprinting analysis and a mathematical formula called the Fourier Transform suggest that the goji is "quite possibly the most nutritionally dense food on the planet!" The book's author, Dr. Earl Mindell (who bills himself as "the world's leading nutritionist"), starts by asking readers how long they want to live. "Eighty years? Ninety? One hundred-plus years? Perhaps even forever?"

The goji not only extends your life, it improves sexual function, helps you see in the dark, alleviates stress, relieves headaches, causes old blood to turn young again—and prevents cancer. To substantiate these claims historically, Mindell trots out someone called Master Li Qing Yuen, "the most famous goji user of all time." Yuen, so the tale spins, was born in 1678 and died in 1930, aged 252. His secret? Daily goji berries.

Mindell prescribes daily glassfuls of his specially formulated Himalayan Goji juice—which retails at a mere forty dollars a liter. Goji berries are certainly good for you as are most fruits. In order to find out how this juice differs from the goji berries available in Chinatown for a couple of dollars a pound, I tried contacting Dr. Earl Mindell repeatedly. None of the messages I left on his answering machine were answered. Calls to FreeLife (manufacturers of his Goji juice) and Momentum Media (the publishers of *Goji*) yielded nothing. According to the National Council Against Health Fraud, Mindell's doctorate is from the University of Beverly Hills, "an unaccredited school which lacks a campus or laboratory facilities." I also tried contacting Mindell at Pacific Western University in Los Angeles, where he is, according to his bio, a professor of nutrition. When I called, the university claimed that, according to their files, no Dr. Mindell had worked there, nor had it ever offered any classes in nutrition.

Venomous debates have sprung up online regarding the juice's putative benefits. Not coincidentally, the most vociferous defenders of the juice often happen to sell it. These salesmen's websites offer goji juice in shipments of four bottles for twelve monthly payments of $186—coming to over $2,200 per year.

Multilevel marketing, or MLM, is a quasi-legal form of pyramid selling, which involves locking recruits into yearly contracts. Repackaging cheap fruit-juice ingredients, and branding them as cure-alls, MLM juice installment plans often leave customers with thousands of dollars in fruit-stained debt, rather than a reprieve from their terminal illness.

Goji juice isn't alone. Tahitian Noni juice nabbed thousands of suckers in the 1990s, until the scheme fell apart. Today, eight bottles of MonaVie açai juice—"the authentic solution for humanity"—sell for $298. Never mind that it's a blend of nineteen fruits (mainly apple juice), or that pure fair trade organic Sambazon açai juice can be purchased at grocery stores for $3.50 a bottle. Yet another addition to the MLM's roster is XanGo mangosteen juice. Its sellers' promotional materials say it heals cancer, depression, fevers, glaucoma, tumors, ulcers, allergies, eczema, thrush, scabies, headaches, back pain, erectile dysfunction, glandular swelling, crooked teeth and on and on. No published clinical trials substantiate these claims, but Oprah loves it. Having generated over 200 million dollars in 2005, XanGo is predicting annual sales of 1 billion dollars by 2009.

Supplements sold through multilevel marketing generate over 4.2 billion dollars annually. The miracle-juice industry is clustered alongside the I-15 in a part of Utah known as Cellulose Valley. Pushing faith is a long-standing Mormon tradition, with generations of door-to-door elders spreading the evangelical message. But there's another reason Utah is the nucleus for these unproven natural remedies.

Senator Orrin Hatch was the architect of the Dietary Supplement Health and Education Act, which absolved producers from having to gain Food and Drug Administration approval before being sold. Not surprisingly, Hatch is himself an investor in Utah herbal companies such as Pharmics and accepts major financial donations from supplement manufacturers. The makers of XanGo made campaign contributions of $46,200 to Hatch in 2006.

Many MLM companies have managed to sidestep regulatory conflicts because unaffiliated salesmen make the fraudulent health claims on their behalf. There doesn't seem to be anything that can be done to prevent families from spending large sums on empty promises: and perhaps these juices, even at their inflated costs, may help certain adherents. The glimmering half-facts of fruit marketing tap into our need to believe. Getting to the truth is its own adventure, as I learned when I started investigating the story of how one truly miraculous fruit was banned in America.

10

Miraculin:
The Story of the Miracle Fruit

There were fruit trees with fruit that sang its way down
dry throats like the gurgle of rippling brooks. . . . Strange
native fruits, flaming with color, bursting with juice.
Nature on holiday, spending herself like a drunken sailor.

—Galbraith Welch, *The Unveiling of Timbuctoo*

M Y DESCENT into Douala, the largest city in Cameroon, is a
dark one. As the plane dips through swirling mist in the
cloud base, only a few streetlights shimmer below, like
smudged pearls in an inky sea of night. A truck with a broken headlight
grins gap-toothedly on the runway.

Cameroon describes itself as "Africa in miniature." With its rain
forests, mountains, deserts, savanna grasslands and ocean coastlands, the
country is a hotbed of biodiversity. Its countless wild plant and animal
species would make it a prime ecotourism destination—if only it were a
little more visitor-friendly.

The transport infrastructure consists of dirt roads that swallow vehicles whole. Stoned teenaged soldiers with Kalashnikovs roam around at
night. Transparency International places Cameroon between Iran and
Pakistan in their list of the world's most corrupt nations. It's hard to get

through a few hours without being embroiled in a bribe scenario. Forget about using credit cards or bank cards: it's cash only, U.S. dollars preferably. Foreigners get fleeced at every turn.

I pay off a guard to get my bags through customs. Outside, a dozen red-eyed young men immediately pounce on me, barking information and hawking services ranging from hotels and taxis to relieving me of my possessions. My contact, a horticultural administrator from Limbe's botanical garden named Joseph Mbelle, is nowhere to be seen. He'd confirmed several times by e-mail that he would be here with a "plycard" bearing my name. Clutching the gift he asked me to bring (a silver-plated watch with "genuine leather strap"), I look around expectantly. At that moment, a dwarf breaks through the throng to assail me. "What company are you with?" he shouts. "What company?"

I shake my head. "I'm not with a company." He leers at me and hobbles away. The back of his T-shirt reads "Ungodly Brutality, Urban Reality."

A man in a fez walks by with a goat. A soldier wearing a red beret starts waving his machine gun at the youths clustered around me. "Danger!" he shouts. "Very dangerous!" As they disperse, the soldier stands next to me, explaining that they want to take my "loot." (I end up paying him fifty dollars in protection money that he starts angrily demanding.)

The Canadian embassy had warned me to stay off the streets at night because of carjacking bandits. Unfortunately, my flight arrived at 2 A.M. Instead of getting a hotel in Douala for the night, I'd been counseled by Joseph to hire him—as well as a car, a driver and a security guard—to meet me at the airport and bring me to the botanical garden that night. An hour and a half later, Joseph, accompanied by two sullen men, finally appears. They'd been pulled over by police and fined for not having the correct paperwork.

As we drive into the darkness, I catch glimpses of shacks and huts in the moonlight. Every fifteen minutes we come to a military roadblock. Invariably, as the gendarmes go through my papers, they try to squeeze payments out of me. One officer asks if it has occurred to me that I have no idea who these men are and that they could be taking me hostage. Shouting matches ensue. As dawn is breaking, we arrive at a crumbling, vine-covered, plaster cube building on the grounds of Limbe's 110-year-old botanical garden. I set up my mosquito tent and fall asleep immediately.

THE FOLLOWING MORNING, I make my way through lush foliage and brilliant sunshine to the entrance of the Mount Cameroon Biodiversity Conservation Center. At the front desk, a painted sign offers visitors "Nature Interpretation for Maximal Enjoyment." The cursive letters recommend hiring a guide: "Otherwise everything will just be green and nice or strange and spectacular—without meaning."

Not wanting everything to be merely strange and spectacular, I take on the services of thirty-seven-year-old Benjamin Jayin Jomi, a colleague of Joseph's. With a sly, baby-faced smile and cocked eyebrow, Benjamin is soft-spoken but has a deep knowledge of Cameroon's plants. As we stroll about the grounds of the living gene bank, he explains how many of these indigenous trees have become raw materials for the international pharmaceutical industry.

The African cherry is marketed in pill form to treat prostate illness. The *Ancistrocladus korupensis,* discovered in 1987 by American bio-prospectors seeking cancer medication in the nearby Korup forest, contains an anti-HIV compound called Michellamine b that is in preclinical development. The yohimbe tree, known as African Viagra, is used in impotency supplements sold with the tagline "killer erections all nite!"

"There is always something new out of Africa," wrote Pliny the Elder in the first century A.D. Today, medical corporations are studying vincamin, the active ingredient in *itongongo* for its effects on hypoglycemia and cerebral metabolism. Benjamin says locals use it the same way nomadic tribes have for millennia: to cure toothaches and facilitate lactation. When rubbed on postpartum mothers' breasts, exudates in these heart-shaped galactagogues increase milk flow.

Cameroonians consume medicinal plants the way Westerners use Advil or Nyquil. The *majaimainjombe,* or blood-of-an-animal plant, is used as a pain reliever. The oil palm counteracts everything from measles to hernias. The bush mango is said to produce Y chromosomes, so members of the Ebu and Bayangi tribes eat it before procreating in order to conceive boys. Benjamin and his wife, Doris, have three children—all girls. Didn't he use bush mango? "Traditions differ," he laughs. "We don't eat it where I come from."

We venture deeper into the wonderland of a garden, passing the remains of a prehistoric asteroid. I pick up a pink, pincushion-shaped monkey fruit with a foamy yellow interior; it tastes like marshmallows. Benjamin points out some wizened, brown, nail-sized pods containing Grains of Paradise, sweet, crispy seeds chewed by parents and then applied to children's sleeping faces to prevent nightmares. I crackle a few kernels between my teeth. They taste sublime—like chocolate-cardamom pellets dipped in clove-infused rose water. In the Middle Ages, European royalty imported these paradise grains by the boatload, considering them to be a literal taste of heaven.

A *Nu Ngan,* or man of the roots, is a traditional doctor specializing in the forest pharmacopoeia. Urbanization has led to impostors capitalizing on the mystery and misinformation surrounding the area's myriad plants. A recent editorial in the *Cameroon Tribune* bemoaned the trend of "doctas" and "gambe men" setting up shop on street corners with herbal cures for any ailment. "What a hoax," it concludes.

While modern-day mountebanks sell magic potions at Yaounde's intersections, many of the garden's potent medicinal plants remain underutilized. The iboga plant bears yellow teardrop fruits sometimes eaten by elephants. More important, its roots and bark are used in initiation rituals by the Bwiti secret society in southern Cameroon and Gabon. Their entheogenic ceremonies, called "breaking open the head," are done to establish communication with ancestors. In addition to being a profound hallucinatory experience, iboga eliminates withdrawal symptoms associated with opiate dependency. Heroin addicts have posted online reports of their habit-kicking trips to Cameroon, where extortionate priests in zombie-white face paint orchestrate nauseating rites lasting up to six days. Iboga detox clinics have sprung up in Canada, Mexico and Europe, although the plant remains banned in the United States.

Leaving the iboga, we come to a shady corner of the garden where another outlawed plant languishes in obscurity. "This is what you came to see," says Benjamin, pointing at a shrub gently sipping bubbly from a nearby stream. The miracle fruit—called the *sweeter* in Pidgin, *assarbah* by the Fante and *Synsepalum dulcificum* by scientists—is the reason I traveled to Africa. Ever since tasting the miracle fruit with Ken Love and Bill Whitman, I've wanted to learn its story. This is where it started. It was in this very garden that David Fairchild first came across it in 1927.

As he wrote in his memoir *Exploring for Plants,* it was so muggy when his ship docked at Limbe (then called Victoria) that walking around was like strolling through a Turkish bath. When his guide at the botanical garden pointed out the *sweeter* plant, he didn't pay too much attention. He ate a few of the berries, which "were not good enough to become excited over, though not at all bad." Shortly thereafter, he was offered a beer to quench his thirst. Taken aback by the sweetness of the beer, he suddenly connected the dots—the miracle fruit! "We at once called for some lemons and sure enough, they tasted sweet as oranges," he wrote. "I had already gathered a quantity of the fruit for seed, but now I stripped the trees."

This little red berry was first mentioned in the eighteenth-century diary of Chevalier des Marchais. The editor who assembled his journals (published as *Voyage du Chevalier des Marchais en Guinea, en 1725*) wrote that, "Chewed without being swallowed, it has the property of sweetening that which one can put afterwards in the mouth which is sour or bitter." (In fact, it doesn't sweeten bitterness—only sourness.) In his 1793 *History of Dahomey,* adventurer and slave trader Archibald Dalzel reported that locals eat it with *guddoe,* stale bread gruel. The first thorough description, by W. F. Daniell in the *Pharmaceutical Journal* in 1852, notes that the fruit is consumed by West Africans before eating a number of native foods—*kankies* (acidulated grain bread), *pitto* (beer) and fermented palm wine—all of which are shockingly sour. He says it also makes unripe fruits taste "as if they had been solely composed of saccharine matter."

As Fairchild wrote, the berry itself isn't very flavorful; the fruit's active ingredient is a glycoprotein called miraculin. The Supermanesque miraculin molecule acts like a key into the lock of your taste buds, a piece of a puzzle that only fits in the presence of acidity. Dr. Linda Bartoshuk, a taste physiologist who studied the miracle fruit for the U.S. Army, explains that there are little sugars attached to the miraculin protein. These sugars position themselves right next to—but just out of reach of—the sweet-sensitive sites on your tongue. The sweet receptors keep trying to get hold of those sugars, almost like a donkey that keeps trying to bite a carrot. But only when you eat a sour food, like a lemon, can the donkey suddenly grab the carrot. When the sugars pop into the sweet receptors, they stimulate a cascade of molecular events resulting in an electrical signal in the nerve. The action potentials propagate along the nerve carrying a message of sweetness to the brain. In other words, the sourness isn't being converted into sweet; its taste is being overwhelmed by the sugars attached to the protein. You don't swallow the sugars attached to miraculin. They linger on your tongue for the next hour, waiting to be reactivated by more sourness. And then the effect dissipates.

After plucking some of the Cameroonian miracle fruits off the garden's trees, I pop one into my mouth, chew it to release a pleasant squirt of nectar, and then swish it about for a moment to allow the juice to seep over my tongue. I spit out the seed, and conduct my own taste test, trying a number of foods before and after. It does nothing for peanuts. A tangerine is barely altered; a grapefruit tastes slightly better. Palm wine is a fermented, effervescent concoction whose pungency is barely masked by the miracle fruit. Its effect with a lemon, however, is so deliriously sweet that it's almost frightening.

There is a deepness to the miraculin flavor that is hard to convey with words. It's a *basso profundo* sensation, like the low frequencies in a symphony. Where at first I could barely lick the puckeringly tart African lemon without wincing, now I'm gulping it down, licking up the juice on my chin. Even the bits on my teeth are ecstatically sweet, like liquefied filaments of pure joy. My head is swimming. Neurons never before activated are firing up my central cortex. I greedily eat up the whole lemon, detecting hints of crystallized grapes and berries.

The American Chemical Society declared it an "unexcelled" sweetener at a national meeting in 1964: "The quality of the miracle-fruit–induced sweetness is more desirable than any of the known natural or synthetic sweeteners." They especially loved it with strawberries: "The delightful flavor of fresh strawberries eaten after miracle fruit is so wonderful that it defies adequate description." Other studies found that it's an all-around flavor enhancer that can transform steak, tomatoes and even certain wines. The media were euphoric: "Babies will go 'bananas,' teenagers will be 'turned-on,' adults will be overwhelmed!"

Entrepreneurs issued patents for its use as an appetite stimulant, appetite suppressant and anorexic agent. Major corporations such as Unilever in the Netherlands and the International Minerals and Chemicals Corporation in Illinois (makers of Ac'cent—MSG) undertook studies, but eventually gave up on the fruit because they found the active ingredient too complex to stabilize and synthesize.

It looked like yet another example of WAWA, an acronym for the region's many business failures: West Africa Wins Again.

IN THE LATE 1960S, a young biomedical visionary named Bob Harvey attended a lecture by Bartoshuk, whose speaking engagements at West Point were classified as top secret by the U.S. Army. "I had long hair and granny glasses and was known to be something of a radical," recalls Bartoshuk.

"I became fascinated," says Harvey. At that point, Harvey was thirty-five years old and his invention of a nuclear-powered artificial heart had earned him millions of dollars. "I guess you could say I was independently wealthy at the time, so I was free to choose what I worked on." He decided to study the miracle fruit under Bartoshuk's guidance.

Convinced of the fruit's potential, he started a company and completed a doctoral dissertation on the fruits titled *Gustatory Studies Relating to Synsepalum Dulcificum (Miracle Fruits) and Neural Coding.* Harvey's research focused on hamsters, measuring changes in the nerves connecting their taste receptors to their brains using a pulse-width modulation system. "It was not a distinguished thesis," says Bartoshuk. "I should've kicked him in the . . . I'd never let a student today get away with what he did, but I was inexperienced because he was my first doctoral student. Anyways, he was more focused on the business end of things."

Where others had failed, Harvey devised a way to make miraculin available in pill form. After biology professors at Florida State University managed to isolate the active protein in 1968, Harvey developed a method of growing the extract in a nutrient-rich culture yielding a hundred-thousand-fold increase. This concentrate was then freeze-dried, crushed into powder and pressed into tablets. The pills were easily stored, distributed and sold, unlike the fresh berries, which bruise easily and have a shelf life of only two days after being picked.

Over the next five years, Harvey raised 7 million dollars from investors like Barclays Bank, Reynolds Metals and Prudential Insurance Company to undertake a major commercial project developing products based on the miracle fruit. By the early 1970s, Harvey's company, Miralin Corporation, was maintaining plantations in Jamaica and Puerto Rico. Their trees were yielding over a million berries a year, and they had employees in West Africa paying children $1.00 per liter for wild miracle fruits, which were shipped home in freezer packs.

Miralin developed a line of sugar-free products—miracle-fruit sodas, miracle-fruit salad dressing and miracle-fruit drops. Their miracle-fruit popsicles were coated in miraculin so that the first few licks would prep the tongue for the sour ice within. In playground tests, these popsicles proved more popular with schoolchildren than those sweetened with sugar. "The results were startling," says Harvey. "It was preferred by a major majority over a sugar-sweetened product."

That never happens.

The miracle fruit was poised to make billions. Diabetics were already gobbling up miraculin. The corporate behemoths in the sugar industry were hell-bent on its demise. Everybody else wanted a piece of the action. LifeSavers was conducting miraculin tests. Warner-Lambert, which manufactured Dentyne, Chiclets and Trident had developed a miracle-fruit chewing gum. Miralin was approached with eight-figure deals.

And then it all turned to dust.

Just as miraculin was about to become widely available, a number of mysterious events took place. "Things started to get very weird," says Harvey. An unidentified car pulled up to Miralin's building and men in sunglasses started snapping cameras in employees' faces to intimidate them. Then, leaving late one evening, Harvey noticed a car idling in front of his office, parked half on the street, and half on the grass. He found it unusual, because everybody in that industrial area had their own parking spot. As he passed the car, he looked in and saw a glowing cigarette tip redden. The car pulled out, following him. Harvey stepped on it. Soon, he found himself in a harrowing ninety-mile-an-hour car chase on winding roads. After one hairpin curve, he yanked on the hand break and came to a stop in some bushes, turning off all his lights. Moments later, the other car flew past him. "Whoever it was, he would've had to change his pants after taking that turn at that speed," says Harvey.

Another night shortly thereafter, heading back to work after eating turkey tetrazzini for dinner, Harvey and his partner Don Emery were taken aback to see the lights on in their second floor offices. Hurrying into the building, they found their alarm system had been deactivated. By the time they ran up to the second floor, the intruder had escaped through the fire exit. Hearing the metal door bang shut at the shipping dock below, they rushed over to the front window, where they saw a car speed off into the night. File cabinet drawers were open, and files were lying scattered on the floor. A police investigation determined that whoever had ransacked their office had done so professionally: none of the locks were jammed and the alarm system had been fully dismantled.

A week later, as the intrigue mounted, what Harvey calls a "cataclysmic event" occurred. On September 19, 1974, Sam Fine of the Food and Drug Administration sent a letter saying that "miracle-fruit products" were not allowed to be sold in any form whatsoever. "Immediately, under penalty of law, I had to close the company down," recalls Harvey.

Up until that point, Miralin had been assured by key government figures that the miracle fruit would easily attain regulatory approval. After all, studies had concluded that miracle fruits were entirely safe even in massive doses. A hundred-thousand-dollar toxicology study undertaken by Miralin demonstrated that rats eating MFC were healthier than those eating pet food. It had no ill effects, even at three thousand times ordinary human consumption.

The reason for the ban remains contentious. Miraculin wasn't a food, ruled the FDA—it was a food additive. Additionally, it didn't qualify for GRAS certification—Generally Recognized as Safe—a system established in 1958 for ensuring food safety. Any food additive in general use before that point, like sugar or salt or artificial grape flavor, required no premarket approval testing. The miracle fruit had been eaten for centuries in Africa, but miraculin had only recently been isolated. The FDA ruled that a food-additive petition would need to be filed, requiring years and millions of dollars of further testing.

This was a period of artificial sweetener fever, with heightened debate surrounding saccharin, cyclamate and aspartame. Their stories shed some light on what Miralin would have faced in attempting to legalize their product.

Following a number of 1960s and 1970s studies linking it to cancer, saccharin was nearly banned by the FDA. Because of consumer protests, it was allowed to stay on the market, sold with labels stating that it "has been determined to cause cancer in laboratory animals" until 2000, at which point a newly Republican Congress repealed the law stipulating that saccharin must carry a health-warning label.

Cyclamate was banned by the FDA in 1969. Since that time, cyclamate manufacturer, Abbott Laboratories, has unsuccessfully petitioned the FDA, at a huge expense. In a strange twist, saccharin is banned in Canada, where cyclamate is legal. Hence Canadian Sweet N' Low and Sugar Twin contain cyclamate, whereas in the States they contain saccharin.

The legalization of aspartame (Equal and NutraSweet), linked to a host of illnesses from impotency to brain cancer, has engendered many conspiracy theories. Donald Rumsfeld was the CEO of G. D. Searle & Company, manufacturer of aspartame, throughout the regulatory period that approved its use in the late 1970s and early 1980s. Thanks to his clout, aspartame finally entered the food stream in 1982. Dr. Arthur Hull Hayes, the head of the FDA responsible for the decision, came under fire for accepting corporate gifts and left his post in 1983 (that year, NutraSweet brought in $336 million dollars). Hayes was promptly hired by Searle's public relations firm, Burson-Marsteller. Concerns about aspartame's toxicity continue to surface in ongoing studies.

With all these suspicious sweeteners swirling around, perhaps the government was just being extra-prudent. FDA employee Virgil Wodicka claims that the agency disapproved of Miralin staging kiddie taste tests before approval had been granted. In another letter, Wodicka cited toxicologists' concerns that children under the influence of miracle fruits might start sipping hydrochloric acid or battery acid. Miralin's executive vice president, Don Emery, has admitted that Miralin's inexperience in dealing with the regulatory commission may have been a problem.

"There's no substance to that charge," counters Harvey. "We had the biggest nutrition and food-additive lawyers in Washington dealing with the FDA. I had a blue ribbon board of directors. This wasn't a bunch of schoolkids—this was serious business. We had an excellent rapport with the FDA." The industrial espionage, the car chase, the break-in—something clandestine was afoot.

Nixon had just resigned, and the stock markets plunged into the worst chaos since 1929. With Washington in a state of upheaval, Harvey couldn't even schedule a hearing with the regulatory commission. His board of directors saw no way forward. They were forced to file for bankruptcy. "To make a real long story real short and sad, I had to collapse everything down and lay everyone off [Miralin had 280 employees] and we all took a huge loss. And in my opinion, so did society."

About five years after Miralin folded, Harvey received a phone call from one of his investors in New York, who worked for a marketing company. He had been reviewing proposals from groups vying to undertake a project for one of his clients when he came across a submission that stood out. This group claimed, on their list of accomplishments, to have been hired by a sugar-industry lobbying group to engineer a strategy that eventually led to the demise of Miralin.

"And they were quite successful," says Harvey, adding that the same group that rubbed out miraculin also purported to have orchestrated the criminalization of cyclamate. When pressed for more information, Harvey says that he has promised to never reveal the source, and that the group "identified their sponsor only as a sugar group representing worldwide sugar interests."

Another source I spoke to, who also asked to remain confidential, asserts that the miracle fruit's demise can be blamed on a competitor. As Miralin was getting ready to launch, Morley Kare, the director of the Monell Chemical Senses Center, was in the final stages of developing sweeteners derived from proteins in two other West African fruits. One was called Monellin, the active ingredient in the serendipity berry. The other was Thaumatin, from the katemfe fruit.

"His motive was financial," says my source. "Kare got the idea that Monell would be damaged in the marketplace, so he contacted the FDA to say that the miracle fruit was toxic. It's an awful story." (It should be stressed that this allegation cannot be validated: Morley Kare died in 1990, and if he did in fact influence the FDA in their decision, no traces are left pointing at his involvement. Adding to the strangeness, Kare patented the use of artificial grape flavor as a bird repellent, making him the inadvertant father of Grapples.)

Harvey, who is familiar with the Kare conspiracy theory, thinks the sugar-industry conspiracy theory is more plausible. "It's all speculative, but the sugar-industry group had more clout and resources to make it happen," says Harvey. "Whoever it was, they convinced someone at the FDA to bypass the due process of law. It was skullduggery. Somebody at some level in the government was paid off."

Lawyers advised Harvey that moving Miralin forward would require years of litigation, an intense emotional commitment with no guarantee of success. "I was all worked up and upset and I was ready to sue everybody and do all kinds of things, but it was really starting to bear down on my health and create other problems," he says. "So my wife and family and I decided we would just move on." He established a cardio-equipment company called Thoratec now worth about a billion dollars. Currently seventy-six years old, he's working on a book about his miracle-fruit experience.

Harvey doesn't believe that the miracle fruit or miraculin will ever become widely available in North America, due to the immense cost of overseeing plantations around the world. "If you translate the tens of millions of dollars from 1968 that's one hundred fifty million dollars today," he says. "Theoretically there's no reason why somebody couldn't. If you ask me what's the likelihood of that happening today, I would say probably not very good. I wouldn't bet on it." Nevertheless, certain developments are afoot that are shifting the odds.

Japan, where square watermelons designed to fit refrigerator shelves and peach-scented pink strawberries sell for vast sums, has embraced the miracle fruit. Researchers have developed genetically modified lettuce and tomatoes containing miraculin; a company named NGK Insulators Ltd. manufactures miraculin pills for diabetics; and miracle-fruit cafés have opened up in Osaka and Tokyo's Ikebukuro district. The cafés' investors developed a way of freeze-drying the miracle fruits so that they can be defrosted and then consumed before sampling a range of sour foods including tart cakes, high-acid fruits, rose-hip teas, lemon gelato and other ice creams. Cake and tea for two comes to about twenty-five dollars, but has one-fifth the calories of a normal dessert.

The Japanese are also at the forefront of research on the lemba fruit. It contains curculin, a protein that, similarly to miraculin, converts sourness to sweetness. The country also sells natural sweeteners such as stevia, katemfe and the serendipity berry that are difficult to find in North America and Europe.

In the absence of miracle-fruit research, scientists at the University of Wisconsin-Madison have patented brazzein, a sweetener found in yet another West African fruit, the ballion. Brazzein is now being genetically spliced into corn. In Gabon, it is called *l'oublie* (the forgetting) because it is so sweet it makes you forget everything.

The miracle fruit is still grown by amateur enthusiasts in a few private gardens in America. From the time he procured a cutting from one of Fairchild's plants in 1952 until his death in 2007, William Whitman ate a miracle fruit every morning before his fruit salad. It's still served before dessert at the Palm Beach Four Seasons Resort. Richard Wilson, of Florida's Excalibur Nurseries, gives many of his berries to cancer patients. Chemotherapy makes foods and drinks taste disagreeable and rubbery. According to Florida oncologists, the miracle fruit appears to replace the persistent metallic chemical taste in patients' mouths with a sweet flavor that allows them to enjoy eating once again. "It also counteracts chemotherapy nausea," contends Wilson.

"There's a lot of uses for it, some that can't be mentioned on camera," Wilson told me when I first interviewed him in Miami. "Just remember: it makes everything sweeter." When, in a subsequent phone conversation, I asked for clarification, he answered forthrightly: "Girls like it a lot because it makes their boyfriends really sweet. They're very candid about why they buy it. One girl said, 'When I suck his dick it's sweet as honey.' I even caught my neighbor robbing berries off my bush at two A.M. I was out there with my pistol and I said, 'What the hell are you doing here?' He said, 'Sorry, my wife just needed those berries right now.'"

Pat "The Blueberry King" Hartmann grew ten thousand miracle-fruit plants in a greenhouse in Michigan in the 1990s. "I monkeyed with 'em," he says. "I thought I was sitting on a gold mine, but it didn't turn out that way." After being told that he couldn't sell the fruits in the United States, he was happy to unload all the plants when a Chinese buyer inquired. "I sold them all to China, five dollars a plant," he says. Hartmann remains incensed about the fruit's unavailability. "The FDA feeds us baloney," he says. "They're dumb. They're dumb. There y'are."

The regulatory limbo is the source of some confusion. One grower said that he'd been told he could sell up to four ounces at a time. "The FDA allows small-scale miracle-fruit operations to grow and sell the fruits," Wilson told me. "From what I understand, the regulations prohibit marketing and selling the fruit on a wide basis."

Another grower, Curtis Mozie, in Fort Lauderdale, says he can sell as many fresh berries as he wants: "It's perfectly okay to sell the fruit. What you can't do is alter the fruit, by extracting miraculin, and then go and sell that." He hasn't ever confirmed this with the FDA, saying he has no reason to call them. He did, however, call me back several times to see if I could figure it out.

I called the FDA about twenty times in attempts to verify the berries' status. Many pointless messages and conversations later, all I could establish with any certainty is that miraculin is banned and the actual miracle fruit occupies a gray area. According to the FDA's 1974 letter, the berries can't be sold either. But today, while the FDA confirms that miraculin isn't on their list of approved food additives ("meaning it isn't allowed to be sold"), they say that the fresh berries aren't covered by their jurisdiction. Fresh fruits, I was told, are regulated by the USDA. The half dozen departments I contacted at the USDA all said that they had no regulations concerning the miracle fruit's sale or use as a fresh food. Then again, none of them had ever heard of the fruit before.

Mozie, who has 1,500 trees, each of which bears hundreds of fruits, sells his harvest for $1.80 a berry at www.miraclefruitman.com. "Eventually, you'll be able to go to a Winn-Dixie or any supermarket and buy my berries," he said. In the spring of 2007, he managed to sell his entire crop—tens of thousands of berries. He had thousands more on back order. "The best thing about the miracle fruit is that it crops year-round," he said, marveling that no one has made money with it yet. "I'm getting orders every day. I'll have a hundred thousand fruits ready to ship by summertime. Think that'll be enough?"

11

Mass Production:
The Geopolitics of Sweetness

Do I dare to eat a peach?

—T. S. Eliot, *The Love Song of J. Alfred Prufrock*

I N THE MIDDLE of the night, on May 16, 2001, a dozen masked
intruders entered a corporate farm in California. Arriving at a test
plot of strawberries, they plunged their hands into the soil and
started wrenching out the plants.

The marauders were a loose-knit band of activists called the Fragaria
Freedom Fighters (*fragaria* being Latin for strawberry). The field laboratory they were stalking through belonged to DNA Plant Technology
Holdings, a bioengineering company developing genetically modified
(GM) fruit. "We invalidated the year's experiment in less than ten minutes, and caused some uncounted amount of economic damage," read a
communiqué issued by the fruit rebels. "A good night's work lying in
shreds behind us, we melted into the night."

This wasn't the strawberry liberation army's first attack. Over fifty
anti-GM actions had already taken place, since the inaugural 1987
demonstration against frost-resistant strawberries sprayed with modified
bacteria. These missions weren't simply frenzies of genetic obliteration.
The Fragaria Freedom Fighters were demanding a return to organic

methods, as a subsequent press release explained. After sabotaging the research field of GM strawberries belonging to another lab, Plant Sciences, Inc., they scattered a variety of organic seeds over the wreckage "to see to it that not only is GM material destroyed, but sustainable agriculture is left in its destructive wake."

Transgenic modification is what happens when DNA is spliced from one species into another. In the past, there was no way a sea urchin could mate with a pussy willow. In recent times, molecular technology has allowed scientists to pursue unlikely genetic combinations, such as goats with spider genes that lactate bulletproof silk, or glow-in-the-dark tobacco leaves bred with fireflies' incandescence genes. In food, this technology has primarily been adopted by large-scale monoculture growers of corn, soybeans and grain crops. It has also been used with sweet fruits, but on a limited scale.

Ice-Minus and Frost Ban strawberries were grown with the addition of genetically modified bacteria called "frost-inhibitors," but they never passed the developmental phase. The Flavr-savr tomato, which contained cold-resistant fish genes (arctic flounder, to be precise), failed after a few years on the market. Most Hawaiian papayas are GM, containing within themselves a vaccination-like genetic dose of the ringspot virus which decimates crops. Papayas' survival has proven to be a mixed blessing for Hawaiian farmers: their fruits are sold in North America, but are banned in many other countries. In 2004, Thai Greenpeace activists dressed in full-body hazmat outfits placed GM papayas into hazardous-waste disposal bins in a widely broadcast action that led to arrests and imprisonment.

It's hard to determine which foods have been altered because they aren't labeled as such. Most consumers reject GM foods when alerted to their presence, but modified crops have creeped into many of the processed foods we eat. With its higher yields, GM has been embraced by industrialized agriculture. Opponents argue that GM crops contaminate nearby fields and warn that not enough testing has been done to establish the safety of molecular engineering. The technology, they say, is to be another step away from diversified and sustainable farming methods. The debate will continue in coming years as transgenic intervention is being touted as the only way to feed a booming global population with diminishing arable land. The technology may also be the solution to a looming banana crisis.

UNTIL THE 1960S, the Gros Michel was the world's top banana. When a lethal fungus called Panama Disease struck, the first reaction was to keep replanting banana trees on new land. After plowing through enormous tracts of virgin rain forest, only to find the virus following wherever they went, growers were finally forced to replace the Gros Michel. Billions of dollars were spent to make the transition to a variety called Cavendish, which has become the main variety in global commerce.

The Cavendish's time draws nigh, however, as it now finds itself under threat by a mutation of the original disease. Entire valleys full of bananas throughout the tropics are succumbing to Panama Disease Race 4 and another highly contagious fungus called black sigatoka. The Cavendish's plight is not unlike what happened during the Irish potato famine, when the country's single variety of potato was struck by a disease that wiped out every plant.

Research teams are working to forestall disaster with the Cavendish: the disease could lead to famine, destabilization and economic collapse in a number of nations. Bananas, an essential part of many tropical countries' diets, are the primary source of carbohydrates throughout East Africa. Seventy-two million metric tons are produced every year in the developing world—compared to a mere one metric ton in the developed world.

To save the Cavendish from viral extinction requires uncovering varieties that are resistant to the new virus. Unfortunately, as the search for these forgotten strains has gotten under way, researchers have discovered that many wild banana varieties have disappeared as a result of logging and mass urbanization. Fortunately, thanks to the efforts of several independent groups, numerous bananas from remote regions have been cataloged and backed up at seed banks around the world. The Thai Banana Club has been rescuing rare bananas from remaining forests and growing them all over the world. Members include Helsinki University's Markku Häkkinen, who lives in an apartment in Finland crammed with weird bananas, and Miami's William O. Lessard, whose nursery has been holding special cultivars in captivity for decades.

According to the U.N., a single banana plant, preserved in a botanical garden in Calcutta, contains DNA that is immune to black sigatoka (it is called *Musa acuminata* spp *burmannicoides*). If other resistant genes cannot be found in bananas, scientists will have two choices: either replace the Cavendish with another variety, or borrow genes from unrelated species. The Laboratory for Tropical Crop Improvement has already created a black sigatoka–resistant banana that contains radish DNA. It seems likely that transgenic splicing will be the only way to save the Cavendish.

Others are suggesting that growers and governments should actually focus on a number of new varieties. Crop diversity could act as a natural buffer against disease. Red bananas and dwarf bananas have started to become available, but alternatives abound, including the blood banana, the sugared-fig banana and the pregnant banana. The Ice Cream banana is silvery-blue on the outside and tastes like vanilla ice cream. The Popoulou's bubble-gum pink interior has a distinct apple flavor. The Haa Haa has bright orange flesh and the Burmese Blue's name says it all. Connoisseurs swear by the Macaboo, otherwise known as the Jamaican Red. In China, the Golden Aromatic is called *Go San Heong* ("You can smell it from over the next mountain"). The Thousand-Fingered is a manic outgrowth of thumb-sized banoonies. The Praying Hands is like a bunch of individual bananas fused into something resembling a baseball glove. However exciting they sound, monoculture farmers aren't likely to focus on these inefficient varieties for the simple reason that growing them goes against the tenets of agrarian capitalism: high yields and reliability. The alarm over genetic engineering isn't too far removed from the suspicion that initially greeted grafting. Indeed, GM could play a vital role in bolstering Africa's food supply. Bananas have been developed that contain vaccines so that children without access to immunization shots can simply eat bananas to protect themselves from deadly viruses. This could prevent millions of deaths every year.

The public, however, is growing wary of the promises of transnational agribusinesses. The green revolution's pesticides and irrigation schemes did lead to increased yields, but it didn't end world hunger; it certainly saved lives, but it also forced farmers to become dependent on chemical seed corporations. The debate surrounding GM must take into account the entire structure of modern food production. Those supporting the technology see it as a way of fortifying a perilous system; those opposing it are calling for sustainable alternatives to petrochemical monocultures. As political philosophers point out, fighting transgenic engineering can be construed, at a certain point, as an act of resistance against the hegemony of global capitalism and its political supplement, liberal democracy.

THE STORY OF the modern banana began in the 1870s, when a twenty-three-year-old American named Minor Keith started building a railroad network through the Costa Rican forest. He planted bananas alongside the railways, not realizing that the seedlings would one day grow into an empire of fruits. Transporting passengers through Costa Rica proved an unwise investment, but Keith managed to turn a profit bringing railcars of bananas to North America. Around the same time, a Massachusetts fisherman named Lorenzo Dow Baker started shipping bananas from Jamaica. They joined forces, and by 1899, the United Fruit Company was overseeing a banana monopoly throughout Central America and the Caribbean.

The "Banana Republics," as they became known, had few other products to export. United Fruit, with their control of countries' banana trade and transportation infrastructures, exerted a dominating influence on local governments. The company took full advantage of its clout, becoming known as "the Octopus" for its many-tentacled agglomeration of political ambition, postal services and banana plantations.

Their human rights violations became legion. For decades, United Fruit resolved labor disputes with armed confrontations. Using brute force as their main negotiating tactic, representatives of the United Fruit Company opened fire on striking workers in the Santa Marta Massacre of Colombia. Their "Great White Fleet" was used in the world wars to transport soldiers and supplies. They funded the Bay of Pigs invasion to protest Castro's nationalization of plantations. They engaged the Honduran army to bulldoze villages to make way for processing plants and banana fields. They were caught by the U.S. Securities and Exchange Commission bribing Central American presidents to reduce export taxes. As papers released by the Library of Congress in 1995 reveal, the company played a decisive role in the military coup that ousted the government of Guatemala in 1954, crippling the nation for decades. They also knowingly exposed workers to lethal levels of pesticides. The use of DBCP, which battles rootworms, resulted in more than thirty thousand South American men becoming sterile. In 1975, the company's CEO, Eli M. Black, committed suicide with a spectacular self-defenestration from his forty-fourth-floor office onto Park Avenue below.

Nowadays, we think of Banana Republic as a clothing chain, but this sordid legacy of exploitation is more than mere dirty laundry. The United Fruit Company, now known as Chiquita, has made efforts to enhance conditions for its workers, building schools and providing health care, and allowing environmental watch groups like the Rainforest Alliance, with their Better Banana program, to recommend adjustments in its growing operations. But in 2007, it was revealed that the company has been funding a Colombian terrorist organization, the paramilitary United Self-Defense Forces of Colombia, to protect their banana plantations.

The company's ongoing political reach was displayed in the trade dispute that erupted between the United States and the European Union in 1999. When the EU imposed a quota system limiting the amount of bananas imported by Chiquita, the American government intervened on their behalf. This might have had something to do with the fact that Chiquita CEO Carl Lindner had become, at the height of the restrictions on his bananas, one of the biggest political donors in America, bestowing more than 5 million dollars. Senate majority leader Bob Dole was given unlimited use of Lindner's private jet. Chiquita's profligacy worked: the U.S. government ended up imposing a 100 percent tax on a number of European products, such as Camembert cheese, cashmere sweaters and luxury handbags. The sanctions served their purpose: the banana quota was lifted and Chiquita emerged with an even greater share of the European market.

Such strong-arm tactics have long dominated international trade. America is the world's leading food exporter, selling 40 billion dollars' worth of agricultural exports a year—funded by 20 billion dollars of taxpayer subsidies. Many of these crops are indigenous to countries that now find it cheaper to buy them from the United States rather than to grow their own. Wheat, for example, originates in Iraq—but now America sells it to them. "For the past year [2006], we've captured almost three-quarters of the Iraqi wheat market, which is quite large, which is well over three million tons," said Bob Riemenschneider, the grain and feed director at the U.S. Foreign Agricultural Services. (During the war, Operation Amber Waves saw the distribution of American-bred wheat seeds to Iraqi farmers; regrowing these seeds is illegal as new seeds must be purchased annually.) It's the same story with corn. On November 5, 1492, members of Christopher Columbus's team in Cuba came across "a sort of grain they called maiz which was well tasted, bak'd, dry'd, and made into flour." Now Cuba—alongside the rest of the free world—buys its GM corn from the United States, although ethanol production is raising prices, exacerbating nutritional problems in poorer countries. This flawed system of Western subsidies has allowed for an ongoing flow of capital into developed countries, while eroding the livelihoods of small domestic and foreign farmers who can't compete with the artificially reduced prices.

This is nothing new: tropical resources have been exploited for the benefit of the North since the mercantilist era. As slaves harvested crops for Europeans, the roots of capitalism and global commerce spread through a damp, dark soil of inequality. Pierre-Joseph Proudhon, an early anarchist, attributed unfairness to private property. Hobbesian types claim it goes back to humankind's ancestral territoriality.

Even today, our food stream relies on migrant laborers who live in subhuman conditions. American fruit pickers aren't farmers or peasants. Many are indentured workers handcuffed by sharecropping agreements. Most of these 1.3 million nomads consider themselves lucky to make minimum wage. They own next to nothing. They have shortened life expectancy. They live in cars, caves and squalid camps full of cardboard tents and plastic sheets. Fruit picking today is "in league with being a rat catcher in Victorian London," says one economist. Yet without these laborers, the world wouldn't eat.

However defective the process, at least southern nations are starting to control exports of their produce to developed nations. Sales of exotics have skyrocketed since the 1990s. According to the U.S. International Trade Commission, mango consumption tripled from 1990 to 2000. Per capita consumption of papayas increased by 56 percent from 1998 to 1999. Nowadays, it's becoming common to see cherimoyas, pummelos, passion fruits and Asian pears in our supermarkets. Unfortunately, many of these are also subpar—despite immaculate appearances.

Selling fruits has become a lucrative and essential form of trade, generating millions of dollars for countries with climates well suited to major crops. The volume of the global fruit trade can't be measured precisely because many countries don't have accurate assessment systems in place. Around five hundred million tons of fruit are produced annually, generating hundreds of billions of dollars worldwide. Retail produce sales approached 55 billion dollars in the United States alone in 2006. The average American spends roughly two hundred dollars on fruit annually. Because of the reversal of seasons, Chile is now the top supplier of fresh summer fruits to North America and Europe from December to May. According to the Chilean Fresh Fruit Association, fruit exports are a major strategic area in the country's development.

Other countries vie to specialize in fruits that thrive in their climes. China sells far more apples than any other country; the United States is a distant second. Turkey sells the most cherries, with the United States again in second place. Belgium is number one in pear exports. India, the birthplace of bananas and mangoes, is the export leader in both those fruits, with their market position set to expand now that the United States is importing Alphonsos. Mexico has avocados locked down, but they've been petitioning the Indian agreement because it will reduce their percentage of the mango pie.

Morocco produces 450,000 metric tons of mandarins compared to China's 11 million, yet Morocco (which sells to Europe and North America) makes over 115 million dollars in mandarin exports, while China only makes 85 million dollars (because it sells to developing nations). To put that in perspective, Spain makes over 1.3 billion dollars a year on its 2 million metric tons of mandarin exports.

With billions flying around in fruit crates, governments have long attempted to export more than they import. This system invariably leads to losers. Because developing nations cannot pay as much for imports as developed countries, the deficits to Northern farmers are covered by enormous subsidies. Export-based economies are also wreaking environmental havoc. Consider this: each year, the United Kingdom exports twenty tons of mineral water to Australia, while simultaneously importing twenty-one tons. Wasteful trade, says the New Economic Foundation, is the rule, not the exception. But as the world's population edges toward 8 billion, the Fragaria Freedom Fighters' proposed return to idyllic organic practices isn't likely to offer any solution to the ongoing tragedy of global hunger.

HUMANKIND HAS NEVER been entirely sure how fruits work. Our ancestors were so awed by the occult power of vegetation that they invented magical ways of growing them. Fragments of esoteric farming manuals abound with spells to create fruits without stones, nuts without shells and fruits without blossoms. Dusty alchemical tomes bristled with directions on sprinkling ox blood around young apple trees so they'd grow red apples or how to pour goat's milk on peach trees so that they'd bear pomegranates. Drilling a hole in a tree and filling it with spices was thought to change the flavor of the fruit. Even Francis "Knowledge Is Power" Bacon believed that watering trees with warm water could induce stoneless fruits. "Of all the signs," he wrote, "there is none more certain or more noble than that taken from fruits."

If the future of fruit farming is uncertain, its past is even more confounding. The fertilizer, pesticide and irrigation for ancient man were prayers, offerings and threats—both to the gods and to the plants themselves. Numerous cultures sacrificed humans in order to ensure crops. Others scared trees into bearing fruits. In Malaysia, sorcerers would strike the trunks of durian trees with a hatchet, saying, "Will you now bear fruit or not? If you do not, I shall fell you." A man in a nearby mangosteen tree would shout back, pretending to be the voice of the durian fruit, "Yes, I will now bear fruit; I beg of you not to fell me."

We didn't really understand fruits, but we knew we could mess with them. We also respected—and feared—their powers. The Indonesian Galelareese tribe believed that anyone who ate a fallen fruit would stumble and fall. Eating two bananas growing from the same bunch, they said, would cause a mother to have twins. In other tribal regions, fruits were ceremonially fed to pregnant women to make trees bear abundant crops.

The complexities of soil management have long baffled us. The Earth, with its life-giving vitality, as well as its deathliness, was a convergence of Eros and Thanatos. Peasant women in Europe used to squirt breast milk on soil to encourage it. We've fertilized with hair, blood, used wigs, rotting blankets and just about anything else that once lived. One Boston grape grower's root system was embedded in the decomposing corpse of a circus elephant. Scientists recently uncovered an entire whale skeleton in the soil of a Tuscan vineyard. As late as 1959, farmers in Tanzania still practiced *wanyambuda,* a fertility rite involving spreading fields with human blood and body parts mashed with seeds. Nowadays, American "residuals management" companies process human waste and sell it in pellet form as fertilizer.

Perhaps the most ghoulish fertilizer ever used on a wide scale was human bones. In the nineteenth century, the frenzy for bones was so acute that British horticulturalists spoke, in all earnestness, of converting "paupers into manure." Tens of thousands of mummies excavated from Egypt were shipped to England to be ground up and spread on fields. German chemist Justus von Liebig, who later introduced chemical fertilizers, accused Britain of grave-robbing from European battlefields to secure their macabre plant food. "In the year 1827," he noted, "the importation of carcasses for manure amounted to forty thousand tons."

At that time, bird droppings—called guano—were, alongside corpses, the most valuable fertilizer around. Large amounts of seagull squiggles were found on small islands near the coast of Peru in the middle of the nineteenth century. The "Guano War" between America and Peru flared up in 1852. At that point, the price of guano hovered around $73 a ton. (As I write this chapter, a barrel of oil costs $72.80.) The Guano Act of 1856 gave U.S. citizens the right to claim uninhabited islands for use in guano extraction. In the 1860s, Spain declared war on Chile and Peru. The shit storm abated only with the advent of artificial nitrogen, phosphorous and potassium.

SHORTLY AFTER MIDNIGHT on December 3, 1984, the leaves started falling from the trees in Bhopal, India. The town's inhabitants, coughing badly, awoke to the sound of neighbors screaming. The streets rapidly filled with people fainting, throwing up frothy blood and having miscarriages. "It felt like somebody had filled our bodies up with red chilis," is how survivor Champa Devi Shukla described it.

The worst chemical disaster ever, the Bhopal tragedy resulted in over ten thousand deaths. Hundreds of thousands were injured. Birth defects, disabilities, diseases and pollution paralyze the region to this day. The ensuing investigation determined that the catastrophe was caused by a leak of methyl isocyanate, an ingredient used to manufacture carbamate pesticides at the nearby Union Carbide factory.

Many of the chemicals used on our fruits were initially developed to be nerve gases and other weapons during World War II. To stay operational, the factories producing wartime chemicals redirected their concoctions into fruit crops. Although they've proven formidable in helping plants overcome adversaries, their use in our food stream continues to be a cause of dissent.

Pesticide manufacturers claim that their products are safe for consumption, but rarely substantiate their claims with any tangible evidence.

On the contrary, independent studies link the chemicals on our fruits to cancers, birth defects, sterility, Parkinsons, asthma, disrupted hormones and a host of other grim diseases. Because these toxins affect human nervous systems and neurochemistry, they are especially unsafe for children, says the National Academy of Sciences. They proliferate in our sperm and ovaries, affecting our reproductive development. Pesticides are found in North American mothers' milk. The *American Journal of Epidemiology* has established a connection between residential pesticide use and breast cancer.

Nobody knows exactly how these chemicals react inside of us because they're often released without any studies. The 1976 Toxic Substances Act stipulates that chemical compounds only need to undergo testing if "evidence for potential harm exists." This is infrequently the case, because most chemicals are released on the market so quickly and with so little public knowledge. All pesticides must be registered with the Environmental Protection Agency, but, as the Environmental Working Group warns, "Because the toxic effects of pesticides are worrisome, not well understood, or in some cases completely unstudied, shoppers are wise to minimize exposure to pesticides whenever possible." Ninety percent of new compounds are granted restriction-free approval. *National Geographic* reports that "only a quarter of the 82,000 chemicals in use in the U.S. have ever been tested for toxicity." There's no transparency with synthetic inputs: they're never listed as ingredients.

It's indisputable that conventional fruits contain traces of hazardous poisons: small doses may be harmless but large amounts are often lethal. From the Bhopal disaster to the military use of herbicides like Agent Orange, agricultural chemicals have taken countless lives.

The names of pesticides like Bravo, Monitor, Champ and Goal have a school-yard innocence that belies these precision snipers' deadly effects. Miticides target mites; herbicides decimate weeds; fungicides murder molds; and rodenticides snuff out small animals. Systemics are chemicals that course through the entire tree: roots, trunk, limbs, branches, sap, flowers, fruits and seeds. Then we eat them. A recent study found thirty-seven different chemicals in a conventional apple. The fruits that retain the most pesticides are those whose rinds or skins are eaten. Conventional strawberries, peaches and raspberries are chemical sponges.

The American Association of Poison Control Centers Toxic Exposure Surveillance System reported 6,442 pesticide exposures in 2003, the majority of which were unintentional. One thousand six hundred ninety-five patients were treated in emergency rooms and 16 fatalities were reported. When patients vomit organophosphate pesticides, ER rooms are instructed to treat the liquid as a "hazardous chemical spill."

Organophosphate pesticides are used on 71.6 percent of apples, 59.6 percent of cherries, 37.2 percent of pears and 27.1 percent of grapes. They break down quickly in the environment, but overexposure can lead to blurred vision, difficulty walking and death. Hydrogen cyanamide is a toxin that can induce nausea, vomiting and parasympathetic hyperactivity. It is used by farmers to achieve uniformity on grapes, cherries, kiwis and other fruits. Methyl parathion is a 1950s neurotoxin that short-circuits insect nervous systems—and can similarly affect humans. Like an edible Taser, it feels like getting hit on your funny bone over and over again, except on your whole body. Until child-safety legislation was passed in 2000, it was all over fruits. Although its use has been diminished, we're still eating it in vegetables, presumably because cooking reduces its toxicity.

It often takes decades of damage before a "miracle plant food" like DDT finally gets phased out. Beaurocracy is indolent by nature; the chemical lobby does its utmost to slow things down even more; and conventional farmers just want to fight off pests as long as they can. Azinphos-methyl (AZM) has been sprayed on apples, blueberries, cherries and pears since the 1950s. "This pesticide has put thousands of workers at risk of serious illness every year," says Erik Nicholson of the United Farmworkers of America. In 2006, the Environmental Protection Agency announced that it would try to phase out the use of AZM by 2010.

Others remain at large. Dieldrin is such a persistent pesticide that some scientists predict it will still exist in the environment when humans are extinct. Methyl bromide causes respiratory illnesses, convulsions and acute mania, alongside intensely depleting ozone levels, yet U.S. agribusiness forces have managed repeatedly to delay its banning, citing emergency "critical uses" stipulations in the Montreal Protocol on Substances that Deplete the Ozone Layer. Supposed to be totally eliminated by 2005, it is still used on strawberries. Mass sprayings of Round Up—Monsanto's glyphosate herbicide—have endangered human health and wiped out myriad flora and fauna. (It is a weapon of choice in the war on drugs, and is used in aerial sprayings to eradicate poppy plantations in

Afghanistan and coca crops in Central America). It is categorized with "noncarcinogenicity for humans" in the United States, despite studies linking it to non-Hodgkin's lymphoma, genetic damage and reproductive problems.

In 2007, a Los Angeles jury awarded 3.2 million dollars to six Nicaraguan farmworkers who claimed they were made sterile as a result of spraying DBCP while working for Dole Food. In similar cases in Nicaraguan courts, Dole and other companies have been ordered to pay more than 600 million dollars to workers allegedly affected by the use of the pesticide, although the companies maintain that the judgment is unenforceable because the law that allowed the workers to bring the lawsuits is unconstitutional.

In Europe, new regulations concerning the Registration, Evaluation, Authorization and Restriction of Chemicals (REACH) are forcing companies to provide data proving the safety of their products. They face stiff resistance from chemical manufacturers. But there should be more transparency regarding the toxins all around us. If manufacturers are so sure there is nothing wrong with genetically modified foods, pesticides and cloned meats, they should have no problems labeling them as such. After all, cancer will kill one in every two men and one in every three women now alive, reports Samuel Epstein, chairman of the Cancer Prevention Coalition.

Like our ancestors, we act in ways that will bemuse future societies. The military-industrial complex lubricates the mass-agriculture system with fossil fuels. Tons of heavy metals and other hazardous, even radioactive, waste is sprayed on American agricultural soil. In the 1990s, manufacturers in Quincy, Washington, intentionally sold toxic industrial waste to farmers as fertilizer, causing an eruption of cancers, brain tumors and pulmonary disease. Farmers in India recently started spraying their cotton and chili fields with Coca-Cola. They say it kills pests just as well as chemicals but costs less. In India, reports Vandana Shiva, tens of thousands of debt-ridden farmers have committed suicide, many by drinking pesticides.

In the late 1990s, scientists investigated the phenomenon of longan trees producing fruits totally out of season. The trees they studied were located near temples where fireworks were used in religious ceremonies. It turns out that gunpowder in the firecrackers induces blossoming. Now most longan plantations use chlorates—gunpowder—as fertilizer.

As a result, residue has already started building up in soils and leaching contamination into nearby ground water. In 1999, forty Thai longan farmers died when their fertilizer warehouse ignited and blew up like a roman candle.

"Seventy-six million Americans get sick and five thousand die from food-borne hazards each year," says the Center for Science in the Public Interest. Though we may think of them as safe, the sad reality is that fruits, like everything else in our industrialized food chain, are a source of food poisonings. Each year, tainted produce is responsible for more illnesses than seafood, poultry, beef or eggs. Even organic produce can leech noxious microbes into our bloodstream, as a recent rash of infected spinach demonstrated.

There are regular outbreaks of E. coli in melons and apple juice; salmonella in orange juice, cantaloupes and tomatoes; hepatitis viruses in strawberries; cyclospora in different berries; and *Vibrio cholerae* in fruits that have been in contact with contaminated water. There have even been documented cases of apple juices containing radioactive isotopes because the concentrate came from orchards within fallout range of Chernobyl.

The central paradox is that fruits seem so pure and healthy, but their dissemination entails so many compromises that by the time we're eating them they've lost the attributes that made them special in the first place.

I can think of no sadder example of our food paradigm than two posters taped to the window of a California IHOP. One is a colorful photo of pancakes heaped with bananas, strawberries, nuts, syrups and whipped cream with the caption, "Welcome to Paradise." Lower down, an 8x10 photocopy states: "Chemicals known to cause cancer or birth defects or other reproductive harm may be present in foods or beverages sold here." Such signs are posted on many fast-food outlets. Heaven isn't a place on earth, at least not at these drive-throughs.

Sustainability is only now starting to be discussed as an objective in agriculture. One innovative program in Europe has been to use government subsidies to pay farm owners to convert their arable land into forests. As a result, European forests have actually expanded by 10 percent in the past twenty years. Such "afforestation" initiatives should be replicated globally. According to a National Research Council report, most U.S. government policies actually "work against environmentally benign practices and the adoption of alternative agriculture systems."

Governmental regulatory boards aren't only beholden to their benefactors—agribusiness corporations have donated over 400 million dollars to U.S. political parties since 1990—they're helmed by industry affiliates with blatantly vested interests.

"More than a hundred representatives from polluting industries occupy key spots at the federal agencies that regulate environmental quality," reports Robert F. Kennedy Jr. The White House office of Environmental Policy chief of staff from 2001 to 2005 was Phillip Cooney, the former chief lobbyist from the American Petroleum Institute. He joined Exxon two days after resigning from his post (following a media scandal about his rewording of documents on global warming). It's not surprising that commercial logging occurs in national forests and parks given the fact that, from 1998 to 2002, the head of the Forest Service was Mark Rey, a former timber-industry lobbyist. The director of the USDA's Food Marketing and Inspection Service was also the president of the National Cattleman's Association. The secretary of agriculture had also been the president of the American Meat Packers Association. In 2006, the head of the FDA, Lester Crawford, pleaded guilty to false reporting and conflict of interest charges for owning shares in the food, beverage and medical device companies he was responsible for regulating.

"Everywhere you look," says Bill Moyers, "the foxes own the chicken coop." Moyers's PBS documentaries on food pesticides revealed that chemical companies have knowingly withheld damaging information about harmful toxins in their products. His research, based on archival evidence from internal company memos, confirms that we live "under a regulatory system designed by the chemical industry itself—one that put profits ahead of safety."

IN 1900, 38 percent of Americans lived as farmers. Today, under 2 percent do so. William Heffernan's *Consolidation in the Food and Agricultural System* reports that about a dozen agribusiness middlemen who package, process and deliver our food (like Cargill, Monsanto and Archer Daniels Midland) make most of the profit whenever consumers buy something produced by farmers.

U.S. farmers commit suicide at four times the national average. In developing nations, workers on gigantic plantations are paid pennies to swing machetes through the sweltering heat, drenched in viscous tree latex and their own sweat, facing lethal levels of pesticide exposure and contending with enormous spiders and scorpions. Epidemiologic studies of field workers show they have increased health problems such as lymphoma, Parkinson's and a gaggle of cancers. With a steady regimen of heat exhaustion, chemical exposure, nerve damage and neurological disorders from backbreaking stoop-picking, farmwork is the most hazardous industry in the United States, according to the National Safety Council.

Workers compensation rates skyrocket when pickers need to climb ladders. As a result, big growers now plant dwarf fruit trees. Agribusiness farms have been trying for years to replace human pickers with robots. Newton Research Labs have developed machine vision systems that can detect different shapes and colors. New Zealand's kiwis are now being sorted, graded and even pollinated by robots that work twenty-four hours a day. These automatons, which are overseen by 1.5 humans per eight-hour shift, "also collect data that will enable coolstore operators to decide which fruit to market and at what time," explains designer Dr. Rory Flemmer. The goal is to have fruit-picking drones whizzing through orchards day and night. While it certainly seems more feasible than the four-thousand-year-old Egyptian method of training monkeys to pick fruits, its merits are debatable. Nobody wants a raspberry harvested by metal fingers.

Mechanical fruit production is almost monstrous. Tree fruits are knocked off their branches by mobile shaking machines that look like Jabberwockies. Their brown metallic pincers clamp around trunks and shake trees so hard that they turn into pointillist blurs. Afterward, pickers walk around and beat the branches with a stick to knock off any remaining fruits. Wind-rowing machines and suction pumps gather the fruits and dump them on a mat. By the end, the battered and abused tree is visibly wilted.

In this era of precision agriculture, electronic devices are used by farmers to monitor their fruits' size, ripeness and firmness. Some growers own satellite systems linked to computers that monitor weather patterns. During ice storms, radar-activated hail guns fire sonic waves that melt the ice pellets into rain. Leonardo da Vinci suggested planting citrus orchards near streams, so that the water could power heat-generating fans in the winter. Today, digital alarms linked to propane heaters and windmill-like hot-air machines protect crops during chills. Helicopters

are flown over plantings to circulate warm air and disperse mist or moisture. Trees are also hosed down by sprinkler systems: as water freezes
onto the plants, it releases heat.

RECENT SURVEYS REVEAL that a significant number of consumers prefer crunchy peaches bred to have a hard, cargo-truck friendly exterior.
Most of us have never tasted a good peach, let alone a downy pastel orb
bursting with sweet nectar when plucked right from the tree. As Marshall McLuhan pointed out, we've become so removed from reality that
we're starting to prefer artificiality. Part of the reason we think fake is fine
is the narrow selection. Turgid peaches aren't even sold in supermarkets,
mainly because they can't be shipped. Ken Slingerland, a peach breeder
for the Ontario Ministry of Agriculture, positively loathes juicy peaches.
"They squish all over your face," he told me. "We believe that consumers
would rather have a peach that's like a nice crunchy apple. Everyone has
their preference, and the texture I like is 'crispy.'" Another farmer I spoke
to referred to industrial peaches as "plastic Kraft dinner fruit created by
dead brains."

Perhaps the crunchers believe the only alternative is mushy, insipid
peaches. But peaches exist that are light-years better than anything we're
being sold. Some growers refer to these as "chiropractic fruits," because
they're so juicy you need to bend over when eating them. But maybe
you'll like crispy best too. As a new peach campaign asks: "Are you a
Cruncher, Leaner, or In-betweener?"

If you haven't, try a real leaner peach before making up your mind. As
David Masumoto, farmer and author of *Epitaph for a Peach,* writes of the
variety grown on his family's land: "Sun Crest is one of the last remaining
truly juicy peaches . . . so juicy that it oozes down your chin. The nectar
explodes in your mouth and the fragrance enchants your nose." Juiciness
doesn't mean the fruit is totally soft; on the contrary, texture is very important. According to stone-fruit expert Andy Mariani, of Andy's Orchard in
Morgan Hill, California, the perfect peach should have "pleasant resistance," a firmness that yields only to sufficient pressure. Only after our
teeth force through the cell walls should the fruit open its floodgates. "It
can be an almost sexual experience for some people," says Mariani.

The texture and flavor of Mariani's Baby Crawford peach is a kaleido-
scope of sweetness, acidity, some astringency—pure peach ecstacy. "It

looks hard, but melts in your mouth. It just oozes nectar," he says. Masumoto, blown away after tasting the Baby Crawford, acknowledges that it's even better than the Sun Crest.

I tasted enough Baby Crawfords to go dizzy when I visited Andy's Orchard in the summer of 2005. The following year, when I called Mariani to check in on the peaches, he sadly informed me that heavy rains and warmer than usual weather had destroyed the entire crop. "The search for the perfect peach is elusive," he said. "It's good for a moment, then a few days later it's gone. It's hard to grow. Nuances in humidity and temperature over one night can drastically affect quality." No wonder growers use any means at their disposal. The fact that fruits ever make it to us is almost heroic.

12

Permanent Global Summertime

They don't have a decent piece of fruit at the supermarket.
The apples are mealy, the oranges are dry. I don't know
what's going on with the papayas!

—Cosmo Kramer, *Seinfeld*

I T'S LIKE A MAFIA convention here," says Jimmy the Greek, my
neighborhood grocer. Taking a sip of steaming coffee from his Styrofoam cup, he points a calloused finger toward the Bentleys,
Hummers and Ferraris in the parking lot near the entrance to the produce wholesaling warehouse. "The owners come to work wearing half a
million dollars. The $300,000 cars, the $150,000 watches, the jewelry, the
rings, the silk suits. I've seen one of them wearing fifteen different
watches—Vacheron, Rolex, Constantin. These guys live in $10 million
homes. You always hear stories about coke parties, whore parties every
night. The money they're making is astronomical, but they're gonna get
burned in the end. You're selling tomatoes—lie low for chrissakes."

I first met Jimmy in front of his Montreal store, where he was
unloading crates of oranges from the back of his van. I started asking
him questions about where he gets his fruit. He was too busy to talk,
but he invited me to join him one morning on his pickups. A week later,
we're shuffling toward this concrete warehouse in the freezing 5 A.M.

darkness. Jimmy shakes his head. "When I got in, I knew there was corruption, but not to this extent. It's sad, 'cause it's *fruits*."

One of the reasons the fruit industry is so lucrative, he says, is that produce is one of the last tax-free businesses. When you go to a grocery store, fruits aren't taxed. The same applies at the wholesale level: everybody pays cash. "They make a lot of legit money, but there's other money there too," says Jimmy. "The business is full of bikers, bookies, loan sharks and gamblers. Drug dealers get into fruits to wash money."

It reminds me of stories I heard in Miami of growers dealing with an Asian organization called the "Fruit Mafia" headed by a woman named Gin Gin. "If you're selling tropical fruits in America, you know Gin Gin," said one longan farmer. In the days when longans weren't widely available, he said, the Fruit Mafia would fly down from New York with three-hundred-thousand-dollars cash in brown paper bags to buy his harvest. Japan's yakuza, Hong Kong's triads and Colombia's drug barons all launder money with fruits. Fruits even figure in the weapons trade. The documentary film *Darwin's Nightmare* features a chilling scene of a pilot revealing how he shipped arms to Angola and returned to Europe with grapes from Johannesburg. "The children of Angola received guns for Christmas," says the pilot, overcome with emotion. "European children received grapes . . . this is business."

The only way to thrive in the fruit racket is to be what the Yiddish call a *hondler*: a fast-talking negotiator who gets things done. Jimmy, with his whip-fast verbal and math skills, is well suited to the industry—even though he despises it. Stocky and ponytailed, Jimmy has angelic blue eyes that contrast with his rugged features, giving him the appearance of an old-looking young person. Because the hours and the pay are like working two jobs at once, he says, the produce business attracts people with gambling debts or drug habits. He signals one of the salesmen and asks him to tell us his schedule. He works Monday through Saturday from 5 P.M. to 6 A.M. Even after punching out he remains on call. "When I'm out of here, customers can still reach me on my cell to place orders 24/7," he says. "I take calls while I'm sleeping. I have to—that's my job." (Shortly after our visit, a forklift clipped him, rupturing his Achilles tendon.)

We enter the wholesaling area, a series of cavernous asphalt rooms filled with crates of fruits. The produce is organized by temperature requirement: some of the chambers are frigid, others are nearly warm.

The different climates are divided by thick strips of plastic curtains that Jimmy pushes through with confidence. They always seem to bounce back at me, pinching and scraping. Honking forklift drivers nearly mow me down on several occasions. "Yeah, watch out for them," says Jimmy. "They're grumpy."

While we walk around the warehouse, Jimmy makes calls to other wholesalers to compare prices. "Everybody's trying to buy low and sell high," explains Jimmy. "It's like the stock exchange." When we arrived, tomatoes were twenty dollars a box. As we stood chatting with a salesman, another shipment came in, and all of a sudden tomatoes were seven dollars a box—priced to move. Jimmy took fifty cases. Spoilage, weather conditions and scarcity account for the constantly fluctuating prices. In the fruit business, the laws of supply and demand are intimately woven into the moods of Mother Nature.

Human nature also plays a big role. "Claiming dumps" is a classic wholesaling trick. It involves bringing in a load of insured fruits. When the load arrives, inspectors who have been paid off declare that the fruit has gone off, meaning the importer can make a claim and get his or her money back. Instead of dumping the fruits, however, or sending them back (by the time they'd return to wherever they came from they'd really be rotten), the wholesaler sells them.

Jimmy explains how easy it is for buyers to "rook" a grower. "It's about building their trust. 'Send me a load,' I pay it. 'Send me another load,' I pay it. Now I have a good reputation, see? Then I tell them I want to make a move. 'Send me ten loads—I'll pay it when they sell.' And then I don't: 'Fuck you.' What's the little Mexican gonna do? Come after me?" Not without a legal division. "Nobody fucks with Chiquita, Del Monte or the big boys," says Jimmy. "They'll come after you."

Wholesaling doesn't favor small growers. Alongside demanding cash up front, farmers can consult the Produce Reporter Company's *Blue Book,* a directory that lists the names of wholesalers, shippers and growers, rating them based on how many times they've shafted others. "Moral Responsibility" ratings determine whether it's worth dealing with the company, and on what terms. The *Blue Book* also mediates and arbitrates in order to help resolve conflicts. It even operates a collecting department that assists in recovering delinquent accounts.

At Hunts Point, New York City's main wholesale market, corrupt USDA inspectors have long accepted bribes from wholesalers. In 1999, an undercover federal investigation code-named "Operation Forbidden Fruit" resulted in eight inspectors and thirteen market employees being arrested, imprisoned or fined. According to the House Agriculture Sub-committee, "The investigation revealed that owners of twelve produce firms at the Hunts Point Market had been routinely paying cash bribes to the USDA inspectors in exchange for lowering the grade of the produce being inspected. This saved the produce wholesalers a substantial amount of money per load, and at the same time defrauded farmers out of tens of millions of dollars."

FROM THE OUTSIDE, there's little indication that the world's biggest wholesale produce market, which feeds over 23 million New Yorkers every day, is anything other than a prison. Coils of barbed wire with grimy plastic bags stuck in them gaze menacingly down from a concrete wall surrounding the perimeter. Hidden away in a forsaken corner of the Bronx, Hunts Point Terminal Market stands among scrap-metal depots, car-crushing junkyards and forklift-repair enterprises. This enormous industrial compound isn't some utopic city of fruit; it's a holding pen for perishable merchandise.

This is where New York's shops go fruit shopping. With its sophisti-cated computerized security system, you can't just walk into the market. Part of the justification for the penal vibe has to do with food safety: it's an obvious target for anyone wanting to contaminate New York's food sup-ply. But rather than feeling secure, as I drive around the market's out-skirts on Food Center Drive, I can't help recalling Jimmy's tales of corruption and crime.

The Mafia connotations aren't only in Jimmy's mind. Shortly after my visit, the New York Police Department broke up a million-dollar gambling ring based in the market. One of the eleven people arrested in "Operation Rotten Apple" was John Caggiano. Owner of C&S Wholesale Produce, Inc., one of Hunts Point's biggest produce wholesalers, Caggiano is also an associate of the Genovese organized crime family, according to Police Commissioner Raymond W. Kelly. "We are determined to rid these mar-kets of any mob activity whatsoever," he announced at a press conference, linking other wholesalers to the Luchese and Bonanno families.

Members of the public aren't allowed past the security gate. Fortu-nately, I've set up an interview with management. After getting clearance from a uniformed official, I enter the colossal factory complex. I drive over some cases of exploded plums rotting on the pavement and park my car. Regulars recommend wearing boots because of all the fruits squishing underfoot. Waste is high because produce won't sell unless it's flawless. At Hunts Point, there's so much garbage it needs to be bulldozed away throughout the day.

The enormous parking lot is filled with hundreds of eighteen-wheelers spewing clouds of smog. At Hunts Point, idling annually generates 32 tons of nitrogen oxides, 31 tons of carbon monoxide and 9.6 tons of soot particles and other volatile compounds that coat the lungs of anyone within breathing distance of the market. Little wonder the South Bronx has one of the highest asthma rates in the United States. Efforts are being made to reduce emissions, but the pollution-enforcement car looks abandoned. It's parked off to the side, badly dented.

I'm greeted at the entrance by Hunts Point's executive director, Myra Gordon. Compact and wiry, Gordon has a tough New York accent that doesn't appear to have mellowed with age. "There are other markets akin to this one," she says, briskly walking me through the debris-strewn market, "but oh so much smaller." Hunts Point generates revenues of over 1.5 billion dollars a year, but it's hard to say exactly how much fruit gets sold here annually. "My calculator doesn't go out that far," says Gordon. "Millions and millions of tons." As Gordon points out managers of different wholesaling departments, I catch glimpses of fruit pallets piled up to the ceiling in temperature-controlled rooms. Groggy foremen shout orders at laborers. Forklifts race around, nipping at my heels. Buyers yell prices and inspect crates of produce.

This main floor is where the sales action happens. The wholesalers' offices are on the second floor, which has the longest corridor I've ever seen. A third of a mile long, the end is so far away I feel like I'm walking into the optical illusion created by two mirrors facing each other. The corridor is narrow; reaching out, I easily run my hand along both walls at the same time. Behind each of the hundreds of doors on either side are people, facts, figures—endless information on the logistics of getting these fruits to market.

Moving into a conference room occupied by several cats, Gordon explains how most of the selling takes place in the middle of the night, so that stores can have fresh fruits ready by the time the rest of the world wakes up. Incoming merchandise can arrive at any time, so the terminal market has to stay open. "It's brutal in here," said Gordon. "You can earn a nice living—but it's blood money. The hours are horrendous. It's very tiring. The profits are nickels and dimes, so you need to be able to stay on top of your business. You have to be singularly oriented. It's very tough at all ends of the spectrum, from the owners down to the guys loading and unloading at the docks."

Over a hundred thousand trucks deliver goods to Hunts Point annually. In the past, rail was the preferred method of shipping, but trains are no longer dependable. Boxcars get waylaid in Chicago shipyards and it takes days to get them back on track, by which point the merchandise has gone off. It takes trains ten to twenty days to get fruits from California to New York. Trucks do it in four.

Even so, it takes weeks for fruits to get to us. After a day of harvesting, precooling and packaging, fruits spend another day in the cooling facility waiting to get shipped out. After being trucked across the country, they sit for an average three days at the wholesaler or in big stores' regional warehouses. Only then does it go on sale to the public. Fruit is usually sold after spending several days on display in the supermarket. It then languishes, on the brink, for a week or more in home refrigerators.

One hundred years ago, Montreal was the North American center for Mediterranean fruits. Cargo ships arrived bearing seventy thousand crates per load. Buyers from Boston, New York and Chicago procured their oranges and lemons at Canadian auctions. The transitions of Montreal's terminal market reflect the changing patterns of produce sales in the twentieth century. In the 1960s, the central market moved from the port to unused land outside the city near major highway arteries and railway tracks. Today, the market is a mere husk. Most of the railway tracks have been paved over and it's now a huge mall full of big box stores like Best Buy, Winners and Costco. Only a few produce wholesalers remain. The market's land has boomed in value, forcing wholesalers to split off and open their own warehouses scattered throughout the city. It's for this reason that there are only a dozen terminal markets left in North America where wholesalers are clustered together.

There has also been a rise in air shipping. High-end fruit with a short shelf life is flown in on large metal containers called LDs—loading devices—that fit into wide-bodied cargo planes. The finest peaches, plums and nectarines sold in New York are packed into LD2s by airfreight consolidators in California or Chile. They're picked up by gourmet wholesalers in New York, such as Baldor's, who get them to upmarket stores where they're on sale the following morning at $7.99 a pound. "Soft, juicy fruit can't survive the cold chain," one New York grocer told me. "If you want quality, it has to be flown in first class."

New York's produce sales began with hawkers and peddlers in Washington Square in the late-eighteenth century. By the late 1960s, the market clogged up the area now occupied by the World Trade Center. Well into the twentieth century, wholesale fruits were sold to the highest bidder. Catalogs were printed and lot numbers assigned to piles of fruit. The auctions were discontinued when growers ended up parting with their crops for a pittance one too many times. An eighteenth-century method in London was to auction fruits "by the candle." Bidding commenced when a pin was inserted into a candlestick near the flaming wick; the winning bid would be the one that came before the wax melted and the pin fell out.

Wholesalers used to specialize in individual fruits, so they'd understand the nuances of their product. As in other industries, produce has seen its share of mergers and acquisitions in recent times. The D'Arrigo Brothers, the biggest wholesalers at Hunts Point, also ship, distribute and are among the largest privately owned growers in the world. Vertical integration and industry-wide consolidating have resulted in a decreased understanding of what it takes to handle fruit properly.

In Europe, farmers associations are countering this trend by marketing fruits based on their regionality. Fruits from specific regions of France are labeled as *Appellation d'origine contrôlée* (AOC). Such designations of origin generate premium prices. Grapes from Moissac and Limousin, peaches from Montreuil, and mirabelle plums from Lorraine are all grown according to ancient methods and are being celebrated for their quality. France's success with geographical indications is spawning a Europe-wide demand for obscure or forgotten varietals with incomparable flavor. "Shelf life used to be everything, but now flavor is being emphasized," says Philippe Stisi, the communication officer for Rungis, the wholesale market on the outskirts of Paris. "We're finding ways to give more pleasure to the consumer."

When I visited Rungis, a vendor told me a joke about a Chinese scientist who invented an apple that tasted like pineapple on one side and like a mango on the other. A rich man, after biting both halves said, "Not bad, but I'll reward you handsomely if you bring me one that tastes like *foufounette* (that means, he said, "*la sexe féminine*"). Months pass, and then the Chinese man returns with a new apple. "Bite this," he says, excitedly, to the wealthy man. The bourgeois takes a bite—and instantly spits it out. "This tastes like shit," he curses. "Yes," says the Chinese scientist excitedly, "now bite the other side!"

That same day, I met a jolly old salesman named Guicheteau hawking crates of apples for fifteen euro. Despite his hard-sell tactics, I declined to purchase them, explaining that I was a journalist, not a retailer. "*Ah.*" He shrugged. "In that case, let me show you something special." From under his table, he took out an apple that had been imprinted with an elaborate, embroidered scene of an upright man lifting a woman onto his phallus. It was like something out of the *Illustrated Kama Sutra*. Seeing that I was taking notes about it, he started playfully calling me names: "*Oh, la vache! P'tit cochon! Salaud! Gourmand!*" (Cow! Piglet! Bastard! Glutton!)

IN AMERICA, fruits are stickered with names, bar codes and places of origin, or branded by lasers that tattoo the information into their peel. These so-called "laser-coded information delivery systems with advanced security clearance" are cleaned, sorted, sized, graded and boxed in packinghouses. Then they're verified by inspectors as conveyer belts move the crates out to the loading dock or to warehouses where they rest in a state of suspended animation before being loaded onto trucks, boats, planes or trains.

Biochemical growth inhibitors and hormone-based retardants have greatly extended the average fruit's life span. Apples can spend close to a year sitting in oxygen- and carbon-dioxide-controlled cold-storage facilities. With an atmosphere similar to Neptune's, these warehouses are the sort of gelid death chambers befitting Walt Disney's head. Soon enough, the fruits are shivering, covered in beads of perspiration and suffering from postcryonic shock. Then we bombard them with ethylene gas, which disrupts their slumber and stuns them into a galvanized ripening process. Bananas release ethylene naturally, which is why other fruits ripen when placed nearby. Supermarket tomatoes taste like cardboard because they're picked green and gassed with ethylene for redness. Many oranges are actually green when ripe, but ethylene disrupts the outer layer of chlorophyll, beckoning the orange color beneath. Alongside getting gassed, many oranges we eat are covered with synthetic dyes.

Dyed oranges used to be individually labeled with purple ink, but such warnings have been dispensed with in North America. Shipping boxes are sometimes labeled, but consumers rarely get to peruse the fine print on orange crates. Citrus Red No. 2 is still used on orange skins in the United States and Canada, despite being banned in Britain, Australia and Norway. The World Health Organization periodically issues warnings, and studies dating back to 1973 have shown it causes internal organ damage and cancer in mice and rats. Cooking with peel, making marmalade, zesting oranges, putting a slice in a drink, even sinking your teeth into an orange may be a flirt with toxicity. In order to establish whether your citrus has been dyed, peel off the rind and look at the white stringy fluff still attached to the fruit—is it orangeish? That's dye that has seeped through the peel into the fruit.

To give our fruits that high sheen and longer shelf life, they are covered with waxes. According to wax producers and exporters Cerexagri, Brogdex and Moore & Munger, there is wax on most of our produce. Some wax comes from shellac (the resinous secretions of small insects) or carnuba (from the leaf of a Brazilian palm tree). But many other fruits are waxed with polyethylene or paraffin wax, by-products of petroleum refinement. We're actually eating oil detritus, what Norman Mailer calls "the excrement of oil." The plastic bags we carry fruit home in are also made from polyethylene. Fossil fuels are used to power tractors and mechanized farming devices, to manufacture the petrochemical fertilizers and pesticides that help fruit grow and to transport fruits from warehouses to supermarkets.

Accordingly, our produce departments look like new-car lots full of enormous, perfect fruits gleaming with wax. The spectrum of colors is heightened by megawatts of directional lighting accentuating the beads of mist dripping from the temperature-controlled display cases. Unfortunately, many of these cars are lemons.

"A goodly apple rotten at the heart, O, what a goodly outside falsehood hath!" as Shakespeare put it in *The Merchant of Venice*. The vainglorious blimps in chain stores look like they're one size too big—plums the size of peaches, peaches the size of grapefruit, mangoes the size of cantaloupes. In Hungary, botoxed North American fruits are described as "unlistenable," like jazz.

In the past, people were skeptical of fruits that looked too perfect: they had "too much of a painted look to be good to eat," wrote one nineteenth-century connoisseur. He was right: around that time, grocers would improve the color of fruits by adding gum arabic, or they'd cover up imperfections with India ink. Calcined magnesia or sulfer dusted onto plums was used to create an aura of unspoiled blooms.

Although the hazards of our supermarket are more insidious today, the words of eighteenth-century Dutch merchant Pieter van der Voort still resonate: he described a bad peach as a "painted whore, because it is nicely and attractively colored, remains hard as an apple, has no taste and yet gives one the eye by hanging around."

"You can't sell a blemished apple in the supermarket, but you can sell a tasteless one, provided it is shiny, smooth, even, uniform and bright," says British author and environmentalist Elspeth Huxley. What most of us don't realize is that you can also return bad fruits. I don't recommend doing this at a farmer's market or with a small independent store. But if I'm at a chain store and I get rooked by some beautiful looking peaches that end up tasting like galoshes, I'll return them. I invariably get my money back, and it sends a signal to the produce manager.

When more than one person asks for a certain variety, the retailer reacts and tries to stock that variety—if it is commercially available. Every five requests are seen as representing the desires of a hundred customers. Demand for quality fruits can help reverse the trend of fruit homogenization

Unfortunately, most us don't know fruit varieties very well, which is the way the industry likes it. Fruits are often sold without being identified by variety: they're just strawberries, not Monarchs, Seascapes or Albions. The strawberries we buy in North America and Europe—firm, red, cold-resistant varieties like Camarosa, Elsanta, Diamante, Ventana—are as reliable as they are flavorless. Little do we realize that there were 1,362 varieties of strawberry described in the 1926 compendium *The Small Fruits of New York*.

There has been a conscious decision by the produce industry to wean shoppers away from varieties. If consumers start learning about all the varieties of fruits, they'll start demanding quality. The recent explosion in new apple varieties has led to an increased understanding of seasonality—and decreased overall apple sales. As Todd Snyder of C&O Nursery explained to me: "You don't increase a fruit's consumption by adding new varieties. When people thought there were only red and yellow apples, they didn't know they were missing out on other varieties. As Galas, Fujis and Jonagolds became available, people all of a sudden became more picky about what they eat."

Supermarkets don't want consumers to focus on varieties because it interferes with their ability to sell low-grade fruits all year long. Even at the height of the season, most supermarkets only stock the same subpar apples, oranges and strawberries. The food industry term for this blurred seasonality is "permanent global summertime." It means that everything is always available—and always mediocre.

Only a tiny fraction of fruit biodiversity is for sale: 90 percent of the foods we eat derive from only thirty plant species. Manifold reasons account for this bottleneck: most fruits aren't dependable. They don't ship well. They aren't precision calibrated. They don't produce anywhere near the volume required by national chains. They may not even produce any fruits some years. They are soft and juicy—they can't be stacked because they'd bruise each other. Certain heirloom fruits are like water balloons. Coe's Golden Drop is a forgotten plum variety that is too juicy to actually be bitten into. You nibble a little hole into it and then suck out its perfumed honey.

Tastewise, some fruits have fared better commercially than others. Satisfactory lemons, oranges, apples, bananas and grapes can certainly be purchased at the local Piggly Wiggly. Others, like strawberries, peaches and figs, can't. It's almost impossible to find an apricot that doesn't taste like coins. Even in Iran, where apricots are called Eggs of the Sun, their bestselling commercial variety is the United States–bred Katy. Fruits need to withstand the rigors of shipping all over the world. Superb varieties such as the white-fleshed Shalah and the red-fleshed Tomcham are only edible a few days after picking. They putrefy fast, they don't look pretty, they have weird ripening demands—but they taste delicious. In short, they're the opposite of shelf-stable fruits.

For the wealthy, the situation is improving; white Angelcots are being sold in high-end New York and California fruit stores at astronomical prices. But for the most part, a good apricot remains an elusive quarry. "In some Persian palace whose quiet garden hears only the tinkle of a fountain it would seem to find its right setting, fitly waiting on a golden dish for some languid Sharazade," wrote Edward A. Bunyard.

IN THE FIRST HALF of the twentieth century, more oranges were eaten than any other fresh fruit. Since concentration was invented, we drink far more than we eat. "The frozen people," as makers of concentrate were known, played a decisive role in lessening the appeal of fresh fruit. They made it easier to whip together some frozen OJ than to squeeze our own. The reason concentrate is always so reassuringly familiar is that it's boiled down, divided into basic components, reconfigured, artificially flavored, and, finally, frozen into cardboard containers. Now we're too busy for that, so we drink Tropicana, thinking it's as good as freshly squeezed.

With urbanization and industrialization, we've become detached from our food sources and have forgotten how things are actually grown. Nonetheless, short of pulling a St. Anthony and moving into a tree trunk, we'll continue to get our food by shopping, rather than foraging or growing our own.

Henry Thoreau saw in fruits "a certain volatile and ethereal quality which represents their highest value, and which cannot be vulgarized, or bought and sold." In many ways, it's true. Raspberries are at their height when eaten straight off the bramble. Within ten minutes of picking them, they've lost something. "That on which commerce seizes is always the very coarsest part of a fruit," wrote Thoreau in *Wild Fruits*. "It is a grand fact that you cannot make the fairer fruits or parts of fruits matters of commerce; that is, you cannot buy the highest use and enjoyment of them. You cannot buy that pleasure which it yields to him who truly plucks it."

Few stores want to deal with the hassle of getting good fruit. Fruits are often located near a store's entrance as a way of luring shoppers in to buy other more expensive processed foods, snacks and soft drinks manufactured by large corporations willing to pay stocking fees. (The ingredient lists on these foods can be unsettling: Kraft's guacamole contains less than 2 percent avocado. Quaker's "Peaches & Cream" oatmeal contains dyed, desiccated apple flakes instead of peaches. Some blueberry waffles are actually blue *apple* waffles. Watermelon Fruit Roll-Ups contain pears, but no watermelon. Some mass-produced fruitcakes are made with turnip.) The reason there's little or no fresh fruit in convenience stores is simple: they aren't profitable enough. As one small greengrocer told me,

"The big stores hate produce. They'd get rid of it in thirty seconds if they could. They're not even making money on it." In that light, it's clear why supermarkets want to stock the cheapest, lowest quality fruits available.

Some progressive outlets make the effort to stock local produce. Even better is shopping at a farmer's market. In a Mediterranean climate, farmer's markets operate year-round, which seems to be the way humans are meant to live. But some fruits are too fragile even for farmer's markets. Visiting a farm is the only way to taste the really good stuff. As nutritionist Marion Nestle writes: "If you haven't tasted fruits that are freshly picked, you have no idea how good they can be."

I've learned the importance of cultivating a relationship with someone working in a quality fruit store, ideally someone who sources local fruits. I make it a point to go see a gnarled, gnomelike fellow at my local market who gives me samples and tells me when to eat what and points out bruised fruits that actually have more sugars. Without someone on the inside, it's a gamble. It's hard to know what's in season, so get to know your greengrocer.

For me, summer means melons, peaches, plums and berries—I don't touch them the rest of the year. (Contra-seasonal fruits sold in the winter from the Southern Hemisphere still aren't tasty enough.) Apples and pears are permanently available, but they're much better in the autumn, as are pomegranates, quinces and persimmons. The quality of citrus soars in the winter. Cherimoyas are late-winter treats, as are papayas—especially with lime and crushed almonds. Spring in Los Angeles announces itself with the arrival of loquats growing on trees all over the city. And then come mangoes, followed by cherries, apricots and fresh strawberries in the late spring. All of this, of course, should be supplemented with local fruits tasted on travels.

PART OF THE REASON we're confused about what to eat is that produce is one of the few areas where we aren't bombarded with advertising. Low profit margins inhibit major campaigns, and without much marketing influence, we buy fruits based on how they look—but as we know, looks can be deceiving. Most customers demand flawlessness, so supermarkets end up wasting anything that isn't impeccable. Twenty-five percent of harvested fruit is said to end up in the garbage. There's a movement afoot in Europe to counter this, where imperfect fruit (what's known as Class 2 fruit in the EU) are sold for 50 pence to a pound less per kilo. Class 2 can be blemished and have prominent defects, but taste just fine—often better than Class 1 fruits.

Choosing fruits at the store remains an arcane science. Consumer habit reports show that the only fruits that we actually put on our shopping lists are apples, bananas, oranges and strawberries—all others are bought spontaneously. If we like the way a fruit looks and feels, we'll buy it. The success of these impulse buys is why peaches and cherries have a better household penetration rate than kiwis or honeydews.

We've evolved countless bizarre methods to supplement basic cosmetic considerations. Marketers call this selection process "supermarket hedonics"—all the squeezing, groping, sniffing, caressing and stroking. Some tests seem to work: flicking the knob off avocados gives a glimpse of its internal state. Others don't: plucking the leaves off pineapples has no bearing on ripeness.

Melon mavens have fine-tuned foibles. Some thump; others leave them out in the sun to sweeten. There are anecdotes of shady growers carving netting onto the surface of melons to lure shoppers. Cantaloupe freaks believe that a more accurate gauge of ripeness is the condition of the indentation where the melon was plucked off the stem. If a stem remains attached, meaning it had to be cut using a knife or scissors, the fruit was probably picked unripe and will never taste good. No stem whatsoever—known as a "full slip" or a "full moon"—is ideal. If bits of the stem are still attached, look for a crack running around the perimeter of the rim. Even if it appears perfect, however, the melon could still taste like potatoes.

Buying fruits at the supermarket is basically a crapshoot because they usually aren't ripe. Fruits are often picked when mature—meaning ready to ship—but they haven't undergone the transformations necessary to reach their ripeness potential. Commercial fruits are picked long before the volatile ethers, converted sugars, softened cells and finely tuned acids have hit their stride.

Citrus, grapes, cherries, strawberries, raspberries, pineapples and watermelon are "nonclimacteric" fruits; they don't continue ripening once they've been picked. They're at their best the minute they're picked, and they go downhill from there. Eat them asap.

Fruits that do continue ripening off the tree are called climacteric. There are gray areas within this category. Apricots, peaches, nectarines, blueberries, plums and certain melons can become softer and juicier off the tree, but their flavor and sweetness doesn't improve once they've been picked. Apples, kiwis, mangoes, papayas and certain other tropical fruits do get sweeter once they've been picked. They convert their inner starch into sugars and actually start breathing heavily as they ripen, giving off all sorts of gases. A banana covered in brown spots is the botanical equivalent of a hyperventilating mother giving birth.

Certain climacteric fruits, such as bananas, avocados and pears, actually need to ripen off the tree—although they still must be picked at the right time. And if they aren't treated properly, they'll never taste right. Bananas turn gray from stints in frigid warehouses and produce sections. Pears are picked using precision calibrated ripeness gauges called penetrometers or potentiometers. After harvest, they then need to be cooled and stored until the optimum moment of edibility. Supermarket pears sit under moldy mist jets for ten weeks in a row, making it hard to find the optimum eating moment. Try buying a dozen pears and eating one a day. If you're lucky, one of the pears will be totally gushing with nectar. "There is only ten minutes in the life of a pear when it is perfect to eat," said Ralph Waldo Emerson.

Penultimate ripeness, when the finicky ethers reach organoleptic gold, when acids and sugars reach the ague of their imbroglio, is fleeting. To recognize this moment requires experience and luck. "The instant of perfection with figs," wrote Colette, is when, "swollen with nocturnal dew, a single tear of delicious gum cries out of its eye." With gooseberries, whether white swans, red champagnes or early green hairys, Bunyard wrote, "the moment of moments and the day of days is on the return from church at 12:30 on a warm July day when the fruit is distinctly warm." All fruits have a peak of deliciousness, but finding that perfection is next to impossible, unless we join the ranks of the fruit hunters.

Part 4

Obsession

Musa paradisiaca, plantain

13

Preservation:
The Passion of the Fruit

Fruit tree, fruit tree,
No one knows you but the rain and the air.
Don't you worry, they'll stand and stare
When you're gone.

—Nick Drake, *Fruit Tree*

A T POVERTY LANE ORCHARDS in New Hampshire, the rainy air is spiced with the wholesome, sour tang of fallen apples. At the back of the farm stand, past the jellies and the syrups, past a counter of unpasteurized, preprohibition-style ciders, sits a display of strange apples billed as "oddball varieties." Passersby are encouraged to sample them.

The rough russet exterior of Ashmead's Kernel, an apple dating back to 1700, conceals a nutmeg-wine flavor. The Calville Blanc d'Hiver, ribbed with lobes like an acorn squash, is a sixteenth-century cooking apple that also tastes superb out of hand. Thomas Jefferson's favorite, the Esopus Spitzenberg is, to my inexperienced palate, the quintessence of appledom. Biting into these apples is like being transported back in time; they're the flavor of life as a Renaissance courtier or a wealthy Virginia landowner.

Self-assured and strong-jawed, Poverty Lane's Stephen Wood credits these oddball heirlooms with keeping his farm operational. From 1965 until the early 1990s, Wood grew McIntoshes and Cortlands—which brought him to the same brink as all apple farmers. "For reasons beyond our control, mainly the global overplanting of apples, it became over-whelmingly clear that the whole industry was circling the drain," sighs Wood, echoing what Gary Snyder told me.

Around the time he realized that buying apples is cheaper than grow-ing them, Wood decided to try something new: old apples. Grafting antique varieties onto his McIntosh trees, he was blown away by some of the flavors he discovered. "It was the picky ones that appealed to me—the ones that can knock your socks off," he says. "Among these very odd vari-eties, we found some that grew to stunningly high standards." Focusing on a dozen main selections, he then set about trying to create a market for them. Packaging them in handsome boxes, he started shipping them to upscale urban markets, where demand is starting to soar. He sells bins of Pomme Grise apples—a variety enjoyed by Louis XIV—for five times the amount his McIntoshes fetch.

"The jury's still out on whether this was wise or not," says Wood, who studied medieval history at Harvard. "It's a wicked risk, but it's better than waiting for a dying industry to revive. I'll tell you in ten years' time whether this was just a little too clever. But it's hopeful and it's given us a chance to grow some really peculiar stuff that might not otherwise be available."

Wood is part of a larger trend. In a market glutted with low-grade fruits, small farmers have had to develop novel and imaginative concepts in order to compete. Those growing unusual, rare and specialty vari-eties—and willing to make an effort to market them—are realizing that they can command a premium price for their fruits.

"For a farmer, sustainability means breaking even," says Jeff Rieger of Penryn Orchard Specialties, a farm located in northern California. "It means making enough to pay taxes and keep the farm going for another year." Rieger's approach has been to focus on a variety of heirlooms: Arkansas Black apples, greengage plums and Charentais melons. He also makes a traditional form of Japanese dried persimmons called hoshi gaki. The fruits are dried and hand-massaged over a period of several weeks, pureeing the interior. As a result, the fruits become sumptuously tender and the exterior gets coated with natural powdery fructose. Rieger sells them for thirty-three dollars a pound at the Santa Monica farmer's market in Los Angeles and to superstar chefs like Thomas Keller. His entire 2006 crop sold out within weeks.

Jim Churchill and Lisa Brenneis run Churchill Orchard. They call it a "rebel brand." They grow the best mandarins I have ever tasted: small, sweet and tart beauties called kishus, an ancient Japanese variety. They got the idea to grow kishus by visiting the citrus variety collection at the University of California at Riverside, which grows more than nine hundred cultivars, including citrangequats, megalolos, orangelos, tangos, citremons, citranges, citrumelos, lemandarins, blood limes, violet-fleshed tangelos and striped green-yellow lemons with pink flesh. "We asked the grad students what they eat because they're out there in those orchards all day long," says Brenneis. The students were unanimous: kishus. Churchill Orchard's kishu harvest gets snapped up at the California farmer's markets and by mail order. Chez Panisse sells them, plain, as dessert. Some of their other fruits are just as unique, like a low-acid vaniglia blood orange that tastes like vanilla cream.

The strangest orange I've ever tasted was on an Ojai farm near Churchill Orchard. It was a nameless mutant that happened to taste precisely like chicken noodle soup. All the components were there: bits of dark meat, white meat, chicken stock and even noodles. But no citrus, in my opinion, can match the deliciousness of the Churchill's kishu. Regulars at the Hollywood farmer's market freak out when they learn that the season is over. "People really into kishus have kishu issues," laughs Brenneis.

For the most part, farmers grow trees bred to produce efficiently and abundantly, rather than the delicious, forgotten anomalies of yesteryear. Those willing to grow fine fruits often care deeply about their crops. Passion is essential: the technical and artistic challenges of a fickle fruit are infinitely complex. It's more lucrative to simply sell one's farmland. Unless, of course, there's nothing else you'd rather be doing.

FOR THE ZEBROFF FAMILY, living off the land requires around-the-clock participation from the whole family, including children, grandchildren, nephews and other relatives. "Farming isn't easy, but we don't consider farming work," says George Zebroff. "We don't work here. We live here."

When I arrive at his farm in British Columbia's Similkameen Valley, in August 2006, his wife, Anna, meets me at the gate. She looks like a grown-up Pippi Longstocking. Adjusting her kerchief as we head into the barn, she offers me a cup of fresh milk with a spoonful of honey. It's symbolic, and delicious: the honey melts into the milk, still warm from the cow. George is busy with a cauldron full of peach jam, so we amble around the rugged land, which seems to be sprouting vegetables, flowers, herbs and fruit trees from every pocket of soil. Their backyard is a sheer mountainside, which creates a perfect microclimate. Chickens wander about—fertilizing the apple trees, says Anna. "Other farms look like Versailles. Here we have weeds around the trees—we leave them be. Our farm is about animals, plants and insects all together." As she speaks, a fly lands on her eyelid and rests there. She's so in tune with nature that she doesn't even seem to notice it.

When he finishes with the jam, George Zebroff comes out to greet us. At first, he's taciturn, almost stern. His wild gray hair is matched by a gnarled beard. He's very tall. Shaking hands with him is like greeting the Colossus of Rhodes. He regards me with suspicion, and asks me again what my book is about. After a brief and pointed inquisition, he approves my topic and becomes quite gregarious, speaking eruditely and with a formal eloquence.

Echoing Wood's assertion that it's actually cheaper to buy food than to produce it, he says that farming used to be highly valued. "Now farmers are viewed as shit disturbers asking for subsidies," he says. "The bottom line is ingrained in our system. The corporate model is predicated on that creed. The malfeasances of corporations are well documented. They are the lifeblood of the world they created for themselves."

As we taste some of their grapes, George sees a sparrow ensnared in some netting on his vines. Gently removing the bird, he launches it into the air. The Zebroffs, unlike many farmers, don't kill any stray animals. Even the snakes that come down over the mountain are trapped and then set free. They bear and share the cost, rather than passing it on to the land, he explains. Their goal was to create harmony instead of making money. The way they did it was to grow many things together. They eat the best, and sell the rest.

He invites me to do the former. Giddily plucking some perfect greengages straight off a tree, I taste them, and can't help noticing a feeling of surrender. What a blessing it is to be alive! Their peaches are sensational.

Ecstatically juicy mulberries stain my fingers. "What we're really after is unstandardized, unhomogenized fruits," explains George. "The integrity of quality is to be seen in their different shapes, textures, flavors, sensations and nutritional values. No two fruits are identical—nor should they be identical."

They planted a wide variety of different trees in different years in order to ensure that they'd have some fruits every year. "Biannual fruit is natural with most trees," he says. "They shouldn't bear every year, so we let them be. Other people force them. They need to take a break. They just knock off and you have to expect it. That in itself is one good reason to grow more than one thing."

Zebroff, whose farm is totally chemical-free, speaks disparagingly about other large organic farmers. He believes that organic should be about diversity—not monocrop farms that still use sprays. The Zebroffs rarely buy anything other than farm equipment, gas and other essentials. They do sometimes shop for food, but only for items they cannot produce themselves. Anna admits to periodically indulging in bananas, but only because they don't grow in this climate.

The Zebroffs say they have no idea if others are living like them; they're too busy looking after the farm to go find out. When I ask if it's still possible for people to live off the land, George answers, "Certainly. If they're willing to do the work. Can anybody do it? Sure they can, but they need to buck a lot of indoctrination. Have you seen the film *City Slicker*?"

"With Billy Crystal?"

"Yes," he says. "Do you know the scene with Jack Palance, where he talks about the secret of life? At one point, Jack Palance lifts up one finger and says: 'The secret is to do only one thing, to have one occupation, to focus and specialize in one thing.' Well, you know what? That's not the secret. That's cultural inculcation." Later, at the end of our interview, when I ask him why this way of farming has been disappearing. Zebroff points upward and reiterates: "Jack Palance's finger."

As we sit on a picnic table, eating some homemade bread and cheese with tomatoes, red peppers and herbs from the garden, I tell George about an exhibition I had just seen at the Vancouver Art Gallery. It was a retrospective of art by the Haida people, a First Nations tribe inhabiting an archipelago sixty miles west of the British Columbian coast. Traditionally, all of their daily objects—canoes, clothes, utensils, oars—were crafted into beautiful art objects. Flipping through one of the books in the gift shop, I came across a Haida saying that had etched itself into my memory banks: "Joy is a well-made object, equaled only to the joy of making it." Zebroff, pouring us a glass of homemade plum mead, nods his head. "That's our creed right there."

WE'RE CURRENTLY LIVING in an era of mass extinction. With the ongoing devastation of the rain forests for timber, paper or cattle ranches, we're losing an estimated 17,500 species every year. Species are disappearing before they've ever been documented.

But extinction is also a natural phenomenon. Ninety-nine point nine percent of species that ever lived are now extinct. Nearly all of them disappeared before humans were even around. We can only imagine what most of those species might have been—but we can also appreciate the abundance that's still all around us.

Farmers help maintain diversity merely by growing certain varieties. David Giordano of Giordano Farms has saved Moorpark apricots by propagating them from the same rootstock his father used when growing them in the 1920s. "You can get Moorparks elsewhere, but they're not like mine," he told me. "I've preserved them." Despite being tinged with green, his Moorparks have an unrivaled flavor. "People roll their nose when they see it," says Giordano. "And then I'll give them a sample and they'll say, 'Oh my God!' It goes from total skepticism to 'How many can I buy?'"

The best way to preserve a fruit is to create a demand for them. After all, fruits want to be eaten in order to propagate. All consumers need to do is want them. Indeed, many fruits believed to be obsolescent are still thriving on small farms. In the 1970s, Kent Whealy of the Seed Savers Exchange—who popularized the idea of heirloom fruits and vegetables—started getting requests for Moon and Stars watermelon seeds, a particularly sweet variety that had been popular in the 1920s. Unfortunately, he couldn't find any. Word got around that the watermelon was extinct. In 1980, a farmer in Macon, Missouri, named Merle Van Doren, contacted Whealy to say he was growing the variety. It has now become one of Seed Savers Exchange's bestselling heirlooms, growing in countless backyards.

The Montreal neighborhood where my mother lives, Notre Dame de Grace, used to be farmland covered in a special type of melon that commanded high prices in fancy New York eateries in the early twentieth century. When the farms were swallowed by residential properties after World War II, the Montreal melon disappeared. Or so everyone thought. Letters were sent to the world's seed banks. In 1996, a package of seeds arrived from the University of Ames, Iowa. They were planted by Windmill Point farm, and a melon patch has been set up behind the Notre Dame de Grace YMCA called the Cantaloupe Garden.

Every year, the International Union for Conservation of Nature and Natural Resources (IUCN) releases their "Red List of Threatened Species." Of the nearly twelve thousand plants that are facing extinction, most are angiosperms, and hence have fruits. Some of the fruits in danger include the *cambucá* from Brazil's Atlantic rain forest, a number of Middle Eastern date palms, five different Turkish pear species and thirty-five species related to the mango.

A group called Renewing America's Food Traditions (RAFT) compiles their own Red List of more than seven hundred uniquely North American plant and animal foods that are at risk of extinction, such as the highly flavored Marshall strawberry and the Seminole pumpkin. It doesn't help that government regulations hamper older varieties' commercial potential. Despite its unique honeylike flavor, the Pitmaston Pineapple Apple is too small to be sold in the EU. In Europe, all seeds that are sold must be part of the registry of seeds, but prohibitive registration fees keep many heirloom varieties from the formal economy. As the *Independent* reports: "It becomes illegal to sell them; so, with no growing plants providing seeds for the future, they're simply becoming extinct."

But the hype over extinct fruits is often overblown. The vanished Taliaferro apple, for example, is often held up as a tragic symbol of what we've lost. Thomas Jefferson described it as his favorite cider apple, yielding a silky champagnelike potion. A little digging, however, reveals that it may still be around.

"I have been seeking it for many decades, as did my father before me," Tom Burford writes in an e-mail exchange. "In the past twenty years, four candidates have emerged that reflect the description in the scanty and conflicting literature about the apple called Taliaferro." It's the Anastasia of apples, with multiple contenders purporting to be heiresses to the tsarist throne of Russia.

"It is agreed among the few that have pursued it so fervently that until it is defined further by perhaps the discovery of a document, it will be difficult to hang the label on any one candidate." In other words, even if it isn't exactly the Taliaferro, there are four cider apples that fit Jefferson's description—and churn out silkalicious cider.

According to the IUCN, *Malus siversii,* the primary ancestor of the cultivated apple, is also facing extinction. But as I found, it has also been protected, despite the pressures on its natural habitat.

THE NATIVE HOME of sweet wild apples is the Tian Shan mountain range in between Kazakhstan, Kyrgyzstan and Xinjiang. The densest concentration of these apples is found outside the town of Almaty, meaning "rich with apple." (It was known as Alma-Ata during the Soviet era, meaning "father of the apple.") Having weathered millennia of changes in weather and pests, these trees carry traits that could be used to prevent a disaster in our food chain. With a population nearing 2 million, the silk route's urban sprawl has been encroaching on these ancient apple forests.

Fortunately, American fruit hunters have spent the past decade combing the wilds of Kazakhstan to collect the genetic pool of apples endemic to the region. Philip Forsline, the curator of the U.S. Department of Agriculture's apple collection in Geneva, New York, oversees 2,500 varieties, including an orchard of primeval apples from Tian Shan. He is certain that the apples he brought back will be used in breeding new and resistant fruits in the years to come.

Geneva is a living museum, with rows upon rows of different apple trees planted side by side. "This is diversity," says Forsline, showing me the collection. "Geneva is Kazakhstan re-created."

Geneva is one of America's twenty-six germplasm repositories. Germplasm is the technical term for active tissue that can be used to grow new plants: seeds, stems, clippings, pollen, scions, cells and DNA. Half a million different plants are backed up by America's National Plant Germplasm System. The seed vault in Fort Collins, Colorado, contains the country's primary agricultural insurance plan: nearly five hundred thousand samples of germplasm, many cryogenically preserved in temperature controlled tanks of -196°C liquid nitrogen.

There are more than 1,400 seed banks around the world. Such institutions, often founded or staffed by amateur enthusiasts, play an enormous role in conservation efforts. Older varieties will be used to breed crops that can keep up with global warming, mutating pests and other threats. As Jack Harlan, a plant hunter who collected more than twelve thousand varieties in over forty-five countries, once wrote, "These resources stand between us and catastrophic starvation on a scale we cannot imagine. In a very real sense, the future of the human race rides on these materials."

The worldwide network of botanical gardens is humanity's effort to preserve our plant heritage. Because many of them are in politically unstable regions, their genetic material is in danger. Iraq's seed bank, which had been located in Abu Ghraib, was demolished during U.S. attacks, but not before two hundred valuable seeds were shipped to Syria for safekeeping. One of the most ambitious conservation endeavors ever attempted by humanity is a seed bank near the North Pole that aims to create a genetic backup of the world's most important plants.

Located inside a permafrost-encrusted cave in a hollowed-out mountain on the frozen island of Spitsbergen, Norway, the Svalbard International Seed Vault is a safety net for humankind's agricultural heritage. In case of a global catastrophe, this glacial Noah's Ark could help re-create the world's crops. Should that fail, an organization called the Alliance to Rescue Civilization is developing a laboratory on the Moon containing the DNA of every life-form on Earth.

UNTIL THE DISSOLUTION of the Soviet Union, the genetic base of pomegranates, which grow wild in the Kopet Dag mountain range east of the Caspian Sea, was centered at an agricultural research station called Garrygala in Turkmenistan. The facility was overseen by Gregory Levin, the world's premiere punicologist. Having traveled to dozens of countries in search of pomegranates, his collection included black, purple and peachy-pink varieties, as well as the seedless Shami, and the super-sweet Saveh, said to be bigger than a baby's head. In the post-Soviet turmoil, Levin was forced to flee Garrygala, minus the 1,127 specimens he spent a lifetime nurturing. Luckily he'd sent backup copies to botanical gardens around the world, so his work has been conserved by, among others, the germplasm repository at the University of California at Davis.

Similarly, when a civil war erupted in Georgia in 1993, the area's seed bank was destroyed, but not before eighty-three-year-old conservationist

Alexey Fogel managed to flee through the Caucasus mountains bearing 226 subtropical fruit samples, including the town of Sochi's complete lemon collection. Others have sacrificed their lives to protect genetic resources. When Nikolai Vavilov's repository was on lockdown during Hitler's siege of Leningrad, peanut specialist Alexander Stchukin and rice collector Dmitri Ivanov chose to perish of starvation rather than eat the precious seeds.

The need to conserve fruits leads people like Maine's John Bunker to roam the countryside looking for vanishing apples such as the Fletcher Sweet, which he found on a tree with only one live branch left. Italian arboreal archaeologists maintain collections of fruits dating back to the Renaissance. There's even a father-son team in Britain who walk along the side of the M1 Highway in order to sample the fruits on trees that sprouted from apples thrown out of car windows. The Brazilian fruit photographer Silvestre Silva spent ten years searching for the white jaboticaba. He finally found a plant near the town of Guararema. Rescued from the brink of extinction, the white jaboticaba has been propagated, and will be bearing fruits sometime in the near future.

Many fruit preservationists have superhero-like alter egos, whether it's "Lemon" Craig Armstrong, "Bananasaurus" Rick Miessau, Franklin "the Fruit Doctor" Laemmlen, Adolf "the Reticent Plantsman" Grimal, Ed "the Mangosteen Man" Kraujalis, or J. S. "the Mayhaw King" Akin. I can almost imagine "Mr. Fertilizer" Don Knipp having a wrestling match against Chiranjit "the Lesser-Known Plant Person" Parmar.

Fruit conservation groups are often informal, grassroots citizen science affairs, like the Society to Save British Fruits, the Association for the Preservation and Promotion of Forgotten Fruits, the Fig Interest Group or the Rare Pit Council.

The Paw Paw Foundation is devoted to the biggest edible tree fruit in North America. Growing from Ontario to Florida, it vaguely resembles a banana, having greenish-yellow-brown skin and a custardy interior. Lewis and Clark survived their trek across the American interior by eating wild "poppaws."

The North American Fruit Explorers (NAFEX) is one of the biggest American amateur fruit associations. The organization started as an alternative to the homogenizing efforts of postwar pomologists. As older varieties were excommunicated from the frigid temples of commerce, a Brotherhood of the Fruit came together.

According to the NAFEX *Handbook for Fruit Explorers,* all members are "devoted to the discovery, cultivation and appreciation of superior varieties of fruits and nuts." They grow Tom Sawyer's favorite apple or the huckleberries Finn was named after. They run backyard research programs. They are fruit pioneers who do it for the spirit of experimentation, not for profit. They don't care what people say about fruits not growing in particular areas—they find ways to make them grow. Professional pomologists mainly work on creating fruits that are aimed at the grocery store cold chain. Not NAFEX—they're amateurs on a quest for excellence in fruits.

"Mysterious and little-known organisms live within walking distance of where you sit," writes E. O. Wilson. NAFEXians are devoted to underappreciated oddities like maypops, a type of passion fruit that bursts when squeezed. The fruits of the chocolate vine, as one member says, look like fat purple bananas that "split open to reveal white flesh and a bead of goop surrounding watermelon-like seeds. What's not to like?"

Just listing some berries will give a rough idea of how many indigenous fruits we've never tasted: crackleberry, whimberry, bababerry, bearberry, salmonberry, raccoon berry, rockberry, honeyberry, nannyberry, white snowberry and berryberry. The dangleberry is a juicy, sweet, blueblack berry. The treacleberry tastes like molasses. The lemonade berry is a Southwestern U.S. berry that Native Americans used to make pink lemonade. The moxieplum is a white berry with a subtle wintergreen flavor. The cloudberry is known as the bakeapple in Atlantic Canada. This name can be attributed to a visiting French botanist. Seeking to identify the fruit, he asked a local what the berry is called, saying, *"la baie—qu'appelle?"* The Newfoundlander thought the Frenchmen was telling *him* the berries' name, so it became known as the bakeapple.

For NAFEXians, it all comes down to flavor. Their handbook explains that most consumers have no idea what they're missing: "This is one of the small tragedies of modern life." As many of them have learned, however, it's a short step from being botanically inclined to becoming a fullblown fruit freak. Die-hard members call themselves "the hard core." As former vice president Ed Fackler once cautioned, *"FRUIT VARIETY COLLECTING IS VERY ADDICTIVE!"*

Members of the California Rare Fruit Growers (CRFG) are also fruit junkies. A national association with thousands of members and chapters all over the country, the CRFG produces that bimonthly bible of fruit fetishism, *Fruit Gardener,* a magazine bursting with luscious full-color fruit porn. The magazine's ads are like something out of the adult classifieds. "Wanted: rare fruits, will pick up, call Mike," or "ATTENTION TROPICAL GUAVA FANCIERS: true INDONESIAN SEEDLESS guava available soon! Preorders now accepted!"

In a recent issue of *Fruit Gardener,* Gerardo Garcia Ramis wrote of people "characterized by an obsession with fruits, a desire to grow 200 species in the backyard, to get every single book with the word 'fruit' printed in it. You know the type. Well, that's me." Spending days going through every back issue of *Fruit Gardener,* I learned that devout members of the CRFG describe themselves as true believers, the neurotic fringe or heavy hobbyists. The more academically inclined among them use the term "fruit votary." A votary is someone devoted to the point of addiction to a particular pursuit. It can also mean a fanatical adherent of a religion or cult.

Many of these true believers aren't comfortable discussing their interest in fruits, as I found out when I called the CRFG's Registrar-Historian. C. Todd Kennedy is referred to variously as a "leading Californian fruit preservationist and historian," a "rare fruit expert," a "renowned fruit connoisseur" and a "fruit rescuer." Excited to speak to him, I quickly understood why the organization also describes him as the "Fruit Crank." He told me that he isn't even passionate about fruits. He only rescues old varieties, he said, out of force of habit. At one point in our conversation, which consisted mainly of monosyllabic answers on his end, he snapped at me. "What were you hoping for—some brilliant breakthrough?" "No, just some stories," I managed, picking up the receiver, which had clattered to the floor in surprise and dismay.

Undeterred, I joined the association and started attending CRFG chapter meetings. One took place in the backyard gazebo of a Victorian mansion near Oxnard, as Mexicans labored in the background. The topic was fruits from South Africa. The half dozen attendees were all octogenarians. "We're basically a bunch of rich old men," one of them joked. Even so, they're also concerned with preservation. As Bill Grimes, the association's president, writes: "We are the instruments for genetic diversity, and the preservation of wonderful cultivars of fruits and vegetables not considered marketable or lacking genetic implants."

The second gathering was held in a San Diego church basement, with about fifty attendees milling around, looking at rare fruits piled on a picnic table. A frail old man approached me as I perused some fruit books on display. He called me "young blood" and said it was important that more young people get interested in fruits, but he understood why it appealed to the elderly. "When you are old and you want something to do with your time," he said, his phlegmatic voice oscillating, "growing fruit trees is a great thing to do."

The evening's main speaker was Dario Grossberger, a specialist in cherimoyas, a fruit from the Andes that Mark Twain described as "deliciousness itself." Beneath their scaly green skin, cherimoyas have a white flesh that tastes like pear cream custard. "Ten years ago, ignorant of cherimoyas, I ate one," said Grossberger. "I loved it. Like the rest of you, when I like something, I plant it. I planted a seed, and five years later there was a great tree. I was very lucky. I called it Fortuna. Little did I know that it rarely works out this easily, as many failed attempts at growing have proven to me since. I was an accidental farmer—now I am president of the Cherimoya Association."

For the next half hour, he explained how to grow and breed cherimoyas. Then everyone crowded around the picnic tables, tasting rare cherimoya cultivars like Coochie Island, Concha Lisa and Big Sister. Noticing a strange yellow fruit sitting in a box off to the side, I picked it up to get a closer look. It was an egg fruit, something I'd tasted with Ken Love in Hawaii. A red-faced, heavyset man immediately raced over. "Put that down," he scolded. "It's a very rare fruit. You're going to smash it!" I tried to apologize, but the egg-fruit man stormed off. For the rest of the evening, he refused to make eye contact. When he cut open the fruit later on, he came up to the group I was standing with and handed out a slice to everyone except me.

14

The Case of the Fruit Detective

As I go back over this list, I see that pale specters of forgotten fruits will haunt my dreams for many years. How can I defend my omissions?

—Edward A. Bunyard, *The Anatomy of Dessert*

A S A CHILD, David Karp loved exploring his grandmother's rhubarb patch. A fruit tree growing in his front yard also fascinated him. Even though it didn't blossom every year, it occasionally burst to life with hauntingly sweet greengage plums. By the time he was a teenager, his preoccupation with fruits took on an odder hue: he started slipping out of bed in the middle of the night and tiptoeing into the pantry to sniff vanilla extract.

As an adult, Karp consummated this infatuation by becoming the Fruit Detective. The moniker came to him in "a moment of levity" after hearing about the movie *Ace Ventura: Pet Detective*. Although his title may be lighthearted, he takes the topic very seriously. Poring through old pomology texts for the names of varieties that were renowned for their flavor, he then spends months or years in search of farmers growing these delicious survivors. Having written dozens of rigorously researched articles about rarities—whether new or nearly extinct—and the maverick specialists who grow them, Karp has become America's authority on fine fruits.

The star reporter and photographer for *Fruit Gardener,* he has also written about fruits for *The New York Times* and *Smithsonian.* Karp often says that most farmers would "sooner raise wombats" than highly flavored fruits. He lionizes those who dare to raise finicky crops, and they love him back because he's just as quixotic as they are. He portrays grower Bill Denevan as having a wild gleam in his eye and quotes him as saying, "I go crazy and do superultra quality."

Karp also goes crazy with superultra quality, trekking around America in search of "exquisite rewards," such as a taste of navel oranges from a padlocked ur-tree, musk strawberries with "an almost mind-bogglingly powerful, primeval aroma" or "fabulously fragrant and expensive" lychee varieties like the Imperial Concubine's Laugh. He sometimes ventures farther afield, having gone to Puerto Rico for mangosteens, Italy for blood oranges and France for greengage plums, but most of his work is done in California.

Growers occasionally provide Karp with items few others might appreciate. One year, Jeff Rieger scraped the pure fructose off some dried hoshi gakis and gave him a little container of the bone-white powder for Christmas. "He went crazy for it," recalls Rieger. "He was like, 'How many persimmons did you denude to get this?'"

Although he can't drink wine because he's a recovering addict, Karp brings a oenophile's appreciation of minutiae to fruits. Robert Parker speaks about Bordeaux wines as "flavor bombs." Karp calls Stanwick white nectarines "atom bombs of flavor." The best fruits, he writes, are characterized by "sky-high levels of sugar balanced by floral acidity" or have an "astonishingly concentrated sweet-tart taste." His articles focus on individual fruits, with him hunting down the greatest examples known to exist—and providing readers with mail-order instructions.

As a self-professed "fruitie," Karp has dedicated himself to the pursuit of flavor, a quest that has become all-encompassing. In one article in the *Los Angeles Times,* Karp wrote that he has at times turned orange from overindulging in apricots. That story ended with him jumping up and down in the dusk, trying to reach a couple of Moorpark apricots that dangled just out of reach. When he finally managed to knock one down, it was, he wrote, a "most transcendent apricot experience." When not out investigating, he wakes up before dawn in order to secure the exact same parking spot in the shadow of the Shangri-La Hotel near the Santa Monica farmer's market. "He's totally nuts," one farmer, a friend of his, told me. "He has a lot of idiosyncrasies, I'm not illusionary about that," says grower Andy Mariani. "Deep down he's a nice guy, but, boy, he's *really* passionate about fruit, to the point of being obsessed."

Although I hadn't been able to interview him for my Hawaiian article about fruit tourism, shortly after his profile appeared in *The New Yorker,* a peculiar sequence of events led to a reconnection with the Fruit Detective. While I was in New York editing some footage Kurt Ossenfort and I had shot of Miami's Rare Fruit Council International, my girlfriend, Liane, was in Los Angeles for a screen test. She ended up going to dinner with Ossenfort's former roommate, Allan Moyle, the director of a film Liane had starred in. Moyle brought along his friend, David Karp.

The following day, Ossenfort was out when the phone rang. Seeing a Los Angeles area code, I picked up the receiver, expecting to hear Liane's voice. It was Karp. When I introduced myself, the first thing he said was, "Your girlfriend's a real hottie." It was rather unexpected, seeing as our only two conversations to that point had been about my interviewing him and us going hunting for cloudberries. Karp called her "the vine," translating her name from French. He was especially taken with Liane saying that cloudberries sound like the sort of thing unicorns eat.

Not long afterward, Liane and I moved to Los Angeles for pilot season. While searching for an apartment, we stayed with Moyle, who owned a multiunit compound in Venice. Filmmakers, seekers, radical activists and other antiestablishment types were constantly dropping in. On any afternoon I might get my inner spirit "defreaked" by a pyramid-building musical healer or go to a group session with Allan's psychic *du jour*—a guy named Darryl Anka from Topanga Canyon, who could enter a trancelike state and channel an all-knowing extraterrestrial force named "Bashar." Moyle himself spoke with a disarming, almost childlike, directness. This quality was intensified by his alopecia, a condition that has left him entirely without hair. Bald and beaming, he brought to mind an impish, oversized baby. He characterized himself as having an insatiable need to be interesting.

One of our neighbors was a film producer named Barota. He was training to become a breatharian. The way it works, he explained, is that you open up your crown chakra and convert sunlight into liquid prana. This nectar is apparently far more nourishing than any earthly food. Explaining that he had some friends who had been breatharians for a year and a half, he pointed out that fruitarianism should be seen as a step toward attaining the goal of human photosynthesis. Although he planned never to eat again, I spied him drinking a glass of wine at a party a few weeks later.

At Moyle's parties, I met someone writing a book on exercises one can do in the bathtub. I met the inventor of a device that zaps you in order to measure your "frequency." I met priests of Eckankar, a new-age group that Allan delighted in calling his cult. One of these priests, before leading us on a group chant, explained that he'd recently dreamed of a fruit that cured him of a mysterious ailment. He'd seen seven doctors, and none of them could diagnose his condition. In his oneiric state, he saw himself eating lots of fruits every day. After starting to eat these fruits, he said, he'd been cured of his ills within three weeks.

At one of Moyle's happenings, David Karp dropped by. It was the first time we'd met in person. He again spoke fondly of Liane, whom he called "the divine Miss Vine." We spoke mainly of writing, and of how little money writers make. He wanted to know how it was possible for a freelancer like me to make ends meet. I explained that it always worked out, somehow. I'd just gotten a two-hundred-dollar check from a magazine for a piece on Werner Herzog. He explained that he received thousands of dollars a month from his "fortunate parent situation" on top of the five-figure advance he'd just received for his book on the fruits of California.

As I started explaining how I hoped to interview him for this book, a hippy-dippy Moyle-type girl interjected, asking if either of us had ever tasted Scuppernong grapes. Of course, answered Karp, and went on to describe their characteristics and origin. "Oh my God, they're the best grapes in the world," she said, fawning. "I can't believe you know about them, nobody knows about them!"

Karp left shortly thereafter. Someone joked that he had run out to chase down some wild berries in the alley. As the party dispersed, I felt like I'd yet again blown my chance to secure an interview.

A YEAR LATER, Liane and I found ourselves living in an apartment in Echo Park that had once belonged to fruit worker rights activist Cesar Chavez. Karp and I stayed in touch, sending brief e-mails and speaking occasionally on the phone. "How's the vine?" he'd ask. He'd speak about himself in the third person: "Fruits like hot summers; Fruit Detectives don't." In one call, he said he was feeling demented because of the medication he was taking for his cluster headaches: "And there's nothing worse than a stupid Fruit Detective."

In an exchange about mangosteens in Montreal, he asked whether they oozed gamboge, a "smegmalike" substance that can perforate the fruits' outer surface. I thought the use of the word smegma in conjunction with a fruit was odd, but he didn't seem to be joking.

Then Moyle and his demure wife, Chiyoko, invited Liane and me to a dinner party with Karp and his girlfriend, Cindy-Cat.* As soon as we arrived, the awkwardness began. I shook Karp's hand and then got caught in a bungled attempt at a greeting with Cindy-Cat, who blocked my forward cheek-kiss movement with her hand, which I ended up shaking. She gave me one of those extremely limp, barely there shakes, and glowered at me. Had I made some social gaffe? I started feeling embarrassed and paranoid. But as they took off their coats, Moyle took me aside. "Did you see how she shook my hand?" he asked. "Like an empress!"

Liane asked her why she was called Cindy-Cat. She didn't reply. After a long pause, Karp said: "Some members of the opposite sex have qualities that are equine, or bovine. In her case they are feline."

"In Liane's case, they're like a vine," I offered, trying to defuse things.

Karp agreed, expounding on her vinelike attributes: "She's so tall . . . slender . . . *blossoming*."

"Creeping," interjected Cindy-Cat.

"Clinging," I added, attempting some "vine" humor.

Moyle then pointed at Cindy-Cat's black-cross medallion and said, "You're so goth." She said nothing. After a moment, Moyle announced that, "Cindy-Cat is not at her best tonight."

The table stiffened. Moyle, who thrives on getting people out of their comfort zone, was relishing the tension. Karp held Cindy-Cat's elbow and said, "You can't make a cat be loquacious when she doesn't want to be." Changing the subject, he started talking about his recently deceased cat, Sahara, whom he missed so much that he still called her name every morning.

*Some friends of hers asked me not to use her real name, so I haven't. "Disappear her," they said, explaining that she is "pathological about being portrayed to the point of being phobic."

We then spoke about Montreal, with Moyle reminiscing about the city he'd left behind since getting kicked out of McGill University. I mentioned that my mother had gone to that same school.

"In the English department?"

"Yes, I think so. Maybe you guys were classmates."

"What was her name?"

"Linda Leith. Do you know her?"

"I fucked her!" he said, laughing uproariously. I started blushing, not entirely sure whether he had or not.

Moyle talked about a party he'd been at with Vladimir Nabokov where a bee was buzzing around annoying everybody. "The bee is indiscreet," he said Nabokov said. The subject then turned to past lives, with Moyle suggesting we all find out who we once were. There was talk of booking a session with the sensitive who helped Shirley Maclaine realize she had been, at different times, a geisha, an orphan raised by elephants and Charlemagne's lover.

Karp explained how, in the past, gentlemen used to bond over fruits the way they do today over golf. As Chiyoko cleared the dishes, he set upon the table some chocolate- and cinnamon-flavored persimmons procured from a "connection" with a scant supply. He sliced them and handed them out. They were incredibly flavorful, chocolaty and spicy. I said that I'd never tasted such delicious persimmons.

Most people have never had a good fruit, he responded, saying how the produce sold in supermarkets is invariably low grade. "Biting into a store-bought persimmon can be like an explosion of dust in your mouth," I said. Karp approved, nodding his head vigorously. "Is that what's called astringency?" I asked, cautiously.

Absolutely, he replied, and then quoted Captain John Smith, who said of the persimmon that, "If it be not ripe it will drawe a man's mouth awrie with much torment." We all laughed. Sensing that I had busted out an appropriate shibboleth, I again asked if I might be able to interview him, perhaps by joining him on one of his research trips. Karp said he'd think about it. He then said we could discuss it if I helped him move some bookshelves the following week. I agreed, excited that I might finally be able to secure my interview.

After Karp and Cindy-Cat left, Liane and I helped clean up. I took a deep breath, trying to assimilate it all. "So have you resolved your karma with Karp now?" Moyle asked.

SOON THEREAFTER, Karp picked me up in Bessie, his white pickup, which also happened to be the name of the white van my punk band had toured in years ago. Karp and I navigated the morass of rush hour on the 405 freeway, making our way to some industrial wasteland to pick up the custom-made bookshelves he had ordered. Karp drove hunched up over the wheel, incapable of focusing both on the road and on the conversation.

Back at his Beverley Hills bungalow, we set up the shelves in his office, a converted garage behind the house. After flipping through *Hogg's Fruit Manual,* an alphabetized, synoptical list of superior fruit varieties, he showed me his library's latest acquisition, an 1880 pamphlet by G. R. Bayley. It contained the line, "Who loves not fruit—ripe, glorious fruit— a priceless boon from the great Creator's hand?" I asked if I could copy it down. He agreed, although he seemed a little ambivalent about it. In exchange, I shared a quote from Dante I'd recently come across about fruits in purgatory.

Like a comic book geek, he showed off his collection of citrus monographs, saying how rare or expensive each one was. Having become something of a fruit nerd myself, I thoroughly enjoyed the time he took to show me his favorite rococo orange paintings. He told me that his two favorite words are *pamplemousse* (grapefruit in French) and *albedo* (the white spongy layer in between a grapefruit's peel and the fruit segments). "I'm fond of saying, 'Look at the *albedo* on that *pamplemousse*,'" he said. I laughed, which made him uncomfortable. I thought it was supposed to be funny; apparently it wasn't.

A small section of his clubhouse sanctum was devoted to the works of Britain's Edward A. Bunyard, whose 1929 masterwork, *The Anatomy of Dessert,* is a fetishistic handbook of fruit exaltation. With each chapter devoted to a different fruit, it is a guide to the complex, highly sweet and highly acidic varieties of the early twentieth century, many of which are barely available today. It's written with an aesthete's ornate flair, as evinced by his description of one plum, the Reine Claude Diaphane: "a slight flush of red and then one looks into the depths of transparent amber as one looks into an opal, uncertain how far the eye can penetrate."

Bunyard's panegyrics have a cult following among fruit votaries. An uptight, eccentric dandy who loved gardening, he wore Le Corbusier glasses, dressed in dapper suits and hated bananas, which he found "negroid." Watermelons were fodder for "the South American negro." In his defense, Karp writes that Bunyard's "contempt for watermelon eaters betray[s] the prejudices of his class." Or perhaps such racism is an echo of internecine protohuman forest warfare.

Like his wealthy father, who bequeathed him a successful nursery, Bunyard fraternized with and was patronized by nobility. When he wrote about the "opulence" of certain fruits, he was being sincere. He considered his garden a collection d'elite—like Jefferson's, minus the slaves. Unfortunately, Bunyard had a family history of fiscal ineptitude, explains Oxford don Edward Wilson, editor of 2007's *The Downright Epicure: Essays on Edward Bunyard.*

Bunyard was a rabid collector whose burgeoning enthusiasms often flopped commercially. This didn't prevent fruit peregrinations to Algiers, South Africa and Tunisia, where he discovered a single apple tree growing "under the shade of Palms in the island of Sfax." He lounged around the Riviera, listening to nightingales and ordering magnums of what his compatriot George Saintsbury called "supernacular" wines.

A bachelor concerned with "the science of living," Bunyard was also a scientist. At the same 1906 conference on hybridization where the word "genetics" was first adopted, he presented a paper on xenia, the effect of pollen on seeds. He spent days dissecting fruit components, using magnifying glasses and other measuring implements to record his findings. He went on what he called "joy rides" of fruit experimentation, planting hundreds of different pear trees, or collecting every specimen of gooseberry known to man. A committed bibliophile, Bunyard's library featured an abnormally large sexology section devoted to erotic mores through the ages. He also wrote about roses, occasionally under the pseudonym Rosine Rosat (which recalls Marcel Duchamp's bicurious alterego, Rrose Sélavy).

Bunyard hung out on the Continent with exiled British homosexuals like Reggie Turner, an aesthete who'd been part of Oscar Wilde's clique. As surviving correspondence testifies, Bunyard was close with the poet Norman Douglas, who also happened to be in what Wilson calls "serious paedophilic" relationships with a fourteen-year-old boy named René Mari who lived in Ventimiglia and a ten-and-a-half-year-old girl named Renata. Wilson ponders—vaguely, given the lacunae—whether Bunyard shared the predilection. This much is certain: Bunyard and his companions not only argued about which ports to pair with plums, they also discussed the proper terminology for smothering a child by lying upon it: Is it "overlaying," or "overlying?"

"Bunny," as Bunyard's buddies called him, sent them sexually candid limericks. In turn, they provided the names of young male secretaries in Florence. Of one, named Parceval, Douglas writes: "He is very obliging. Don't hesitate to use, and abuse, him!"

Bunny seemed more tormented by melons than Italian boys, bemoaning the unattainable ideal of melonic perfection. Depressed and bankrupt, he shot himself in 1939, at the onset of winter, postponing his suicide until the end of cantaloupe season. "Into the last shadows let us not follow him; we know not what they concealed, or in what secret temple his soul may have found its peace," read his obituary. "All the time he is searching, seeking. He is looking for the nurseries of Heaven."

Like many of his disciples, Bunyard took a bestial delight in produce, writing of his quiet carnal anticipation for the "lustiness" of peaches. He composed wistful passages about "the plump turgescence of youth," and in pears found the "soft rapture of attainment."

Other fruit writers have also dropped libidinal hints. The nineteenth-century clergyman and novelist Edward P. Roe noted how strawberries—"morsels more delicious even than 'sin under the tongue'"—directly affect "that imperious nether organ which has never lost its power over heart and brain." Lee Reich, author of *Uncommon Fruits for Every Garden,* channels Humbert Humbert when discussing an odd little Yugoslavian pearlike fruit called the Shipova: "Shipova. Speak the name. It makes a pleasant sound, especially with the middle syllable emphasized and drawn out ever so slightly. Intoning the name is equally pleasant to the lips, the end of each syllable leaving the lips poised for the beginning of the next: Shi . . . po . . . va."

Karp moved to his filing cabinet and pulled out some "triple-X" photos he'd taken of golden raspberries. He said he wouldn't be surprised if they were banned upon publication. He then confided that whenever Cindy-Cat wants to get him really excited, all she has to do is mention a certain variety of plum, *Prunus subcordata,* if I recall correctly.

There's a subtle connection between fruit sublimation and another inexhaustible foible. Bruce Chatwin, in *Utz,* tells of a concerned mother bringing her porcelain-obsessed son to the doctor.

"'What,' Utz's mother asked the family physician, 'Is this mania of Kaspar's for porcelain?'

'A perversion,' he answered. 'Same as any other.' "

But behind every fetish lies a more complicated need. Emperor Rudolf II was a devout naturalist who owned a coco-de-mer and a portrait, painted by Arcimboldo, that depicted his head as consisting entirely of fruits: a pear nose, apple cheeks, mulberry eyes and grape-pomegranate-cherry hair. Collecting exotica, he said, was the only thing that alleviated his depression. Like Augustus the Strong, he became impassioned by the fever of acquiring porcelain, emptying his coffers for figurines of bejeweled canaries in pearl headdresses, or dancing naiads as pretty as Rodin's marble sirens, or melancholy princes with slender fingers and gilded epaulets reclining on hoofed thrones. Such collectors have a drive for completeness that is, at its core, a pursuit of perfection, of death. "The craving for porcelain is like a craving for oranges," Augustus remarked at one point, trying to explain his "porcelain sickness."

There are certainly overlaps. Both fruits and ceramics were sources of wonder, tokens of status, emblems of esteem. On a surface-level, they suggested sex and luxury. Beyond the concupiscence, however, was something more insubstantial. "The search for porcelain," concludes Chatwin, is a quest "to find the substance of immortality."

Turning back to the books, Karp explained how, of all fruit writers, Bunyard is of paramount importance. Karp's own soon-to-be-released guide to thirty-five fruit varieties is described as "a book of fruit connoisseurship based on the principles elucidated by Edward A. Bunyard." Bunyard is also a touchstone for Karp's other gurus, such as Andy Mariani and the "Fruit Crank," C. Todd Kennedy, who compares a nectarine's ideal flavor to a pheasant hung until high. While I was trying to make a joke about the taste of partridge in a pear tree, Karp pointed at a book called *Uncommon Fruits and Vegetables* and said that the author, Elizabeth Schneider, was his mentor. "She was my *inspiratrice*. She showed me that it is possible to take produce very seriously."

I wondered if Karp, by telling me of his own discipleships, was insinuating that he wanted to, in turn, impart some of his knowledge to me. Trying to steer the conversation toward setting up a proper interview, I asked if he had any fruit escapades planned in the near future. Yes, he said, he was planning a trip to Ventura County for strawberries. I asked if I might accompany him in order to write about his work. He said he didn't see why not, provided I'd hold his reflective screen while he took photographs. Heading back home, I felt elated. My interview was in place.

THE DAY BEFORE our expedition, Karp e-mailed me an itinerary with a 4 A.M. departure time. Even though our destination was only an hour and a half away, he wanted to start shooting at dawn. He suggested I sleep in his guest room so that we'd be able to leave without delay the following morning. On March 9, 2005, Liane dropped me off at the Fruit Detective's home with some pajamas, a change of clothes, a toothbrush—and a notebook.

From the moment I walked in, Karp was eyeing my white running shoes. I asked if I should remove them, and Cindy-Cat said no. They talked about the living room rug, emblazoned with pomegranates. It had belonged to Karp's recently deceased mother. They also explained that they'd been having termite problems. Karp suggested that they get an aardvark to deal with it. (He loves anteaters so much that he once climbed through the bars at the Philadelphia zoo in order to interact with one.)

After another minute or so of small talk, during which Karp couldn't take his eyes off my shoes, he finally asked if I had brought any boots. I said no. "Those are totally unpractical," he frowned, flustered. "If you are going to be a deputy Fruit Detective, you need proper footwear for field-work."

Luckily, he had an old pair of Fruit Detective boots lying around. They were the right size. Crisis averted, we turned in. As Karp was brushing his teeth, Cindy-Cat took me aside and said she was worried about Karp's driving. "If it seems like he doesn't see a red light, ask him," she said. "Make sure he sees it, because sometimes he doesn't." I assured her that I would, and that everything would be okay. She didn't seem convinced.

KARP WOKE ME up in the darkness. We ate a few dried persimmon slices and hit the road. To be more accurate, we lurched forward in fitful starts and stops. Karp drove as if the asphalt was his enemy, jabbing at the gas pedal and the brakes like a tap-dancing circus bear. We barely spoke for the first forty-five minutes as he wrestled with Sunset Boulevard.

On the open highway, he started telling me a little about his strawberry research. He'd been working on this story for three years, and had inter-viewed dozens of sources, amassing more than 150 pages of notes. I asked

him which strawberry he liked best. He started telling me about one particularly delicious variety called the Marshall, which had disappeared
except for one plant that he'd found growing in a seed bank. As I wrote
down the word "Marshall," Karp jammed the brakes and asked me what I
was doing.

"Umm, just taking notes. For our interview."

"I feel very uncomfortable about that," he replied. "I don't want to have
to watch what I'm saying around you, and censor my thoughts. What if I
were to say something off color? What if I said the word 'nigger'?"

I put away my pen and we drove on in silence.

AN HOUR LATER, we arrived at Harry's Berry farm in Oxnard. Walking
through the fields, I was so excited to be surrounded by strawberries that
I immediately, and surreptitiously, started eating them. Their two varieties, Gaviota and Seascape, were juicy, sweet and cool in the morning air.
They grew on little plants close to the ground in neat rows stretching into
the distance, where laborers in hoodies were stooping and picking in the
trenches.

Karp wasn't that interested in tasting the berries. He was more concerned with the light changing as an overcast dawn appeared. He seemed
tense. I held the reflective disk as he snapped some photographs. All of a
sudden he became agitated. "Did you see that? Did you see how the sky
just opened and the light changed?" He started shooting with machine
gun ferocity.

It was like watching Austin Powers shoot fashion models. He was taking dozens of shots a second, oohing and aahing, cooing, "Spectacular—
oh yeah, oh YES! I'm going to open this up all the way—ooh!" I
commented on it: "So that's why they call it food porn." He stopped
shooting, became dead serious, and said, "What do you mean?" I tried to
explain myself, but he stopped listening and went back to the glistening
strawberries.

The growers, Rick Gean and Molly Iwamoto Gean (whose father was
the Harry in Harry's Berries), came over to greet us. They asked how our
drive had been. Karp answered that it hadn't been too hairy.

Rick asked if I was a budding fruit grower. "No, I'm a journalist working on a story about people who are passionate about fruits," I replied.

"Well, you're certainly with the right person." He laughed.

"Adam's a fruit groupie," said Karp, not laughing. "He's just here to help me take the photographs."

I looked at him in amazement. Yes, I'd agreed to hold the reflector, but he'd also agreed to be interviewed. As we sat down in their kitchen, Karp began grilling Rick and Molly, the way a real police officer—or detective—might. I was sitting in between them, turning my head to watch like a tennis spectator as they discussed sprays, growing methods and strawberry varieties. Karp asked whether they might consider growing the Marshall. He was very curious to find out whether flavor was their primary objective. Molly said that flavor was relatively important to them, but so were yields. He shook his head and pursed his lips as though confirming an incriminating piece of evidence. Feeling increasingly weird about sitting there, and not daring to take any notes, I excused myself and walked back to the strawberry fields. After eating some berries and looking up at the sky to see if I could notice any change, I wrote down what had just transpired.

When Karp was finished, we ate lunch at a small diner with old citrus crate art on the wall. When his sandwich arrived, he took out a couple of containers of dried hot peppers and poured copious amounts between the buns. I remembered that Ossenfort had once mentioned Karp's addiction to capsaicin, the active ingredient in chili peppers.

We hardly spoke on the way home, Karp being so focused on his non-hairy driving. I started to nap but he woke me up. I made small talk about traveling to different countries to find fruits. He said that he wasn't into traveling anywhere except to some archipelago in between Madagascar and Perth in the Indian Ocean. "It's a perfect sixty degrees three hundred sixty-five days a year," he said. "There are no trees or fruit because it's too windy, but you can walk along a desolate lunar landscape and the best part is that there are no other humans anywhere nearby to ruin it."

ABOUT A MONTH LATER, I received a mass e-mail from Karp addressed to the many breeders, growers, marketers and nurserymen he consulted while researching his strawberry article.

He drew a comparison between writing about fruit and growing fruit: "one can put all one's energy and skill into raising a superior product, only to be confounded by forces beyond one's control." The inhibiting force in question was space constraints. Nearly half of his article had been amputated by editors. What remained, he said, was "a barely adequate summary, with none of the complexities" that he'd spent more than three years gathering.

A few weeks later, as I was getting ready to head back to Montreal, Karp invited the vine and me to brunch. We brought our friend Sarah, a writer and record-store employee who was visiting from San Diego. We arrived late because some vandals had thrown a cinder block through our rear windshield the night before. Cindy-Cat was making pancakes with black raspberry syrup.

"I love black raspberries," said Sarah.

"Oh?" asked Karp, "have you ever tasted black raspberries?"

"I think so, in Sweden."

Karp started tweaking out, barking that the fruit's native range is actually in the Pacific Northwest and that she never could've had a proper black raspberry anywhere in Scandinavia.

I told him that I'd tried one of his favorite fruits, the white apricot. He said that it wasn't enough to taste a few fruits. To be able to write about any fruit properly, he said, his voice rising, you have to spend years and years going to private farms and to corporate growing facilities, to germplasm repositories and to university labs and you have to keep on going over and over and over again.

15

Making Contact with
the Otherworld

And then he went into his pocket and took out
a seed for a tree . . . He put it in my hand and
he said, "Escape—while you still can."

—*My Dinner with Andre*

WHEN I CALLED Kurt Ossenfort to fill him in on my progress with the Fruit Detective, he asked if Karp had told me about the Children of Light, a group of immortality-seeking virgins and eunuchs growing heirloom date varieties in the Arizona desert. I had heard of individual fruitarians wanting to live forever, but this was something even stranger: a utopian fruit-growing cult.

I asked Karp about the Children of Light, but he was reluctant to discuss them. After our last outing, it was clear that we wouldn't be visiting them together, but I decided to go see them anyway—before it was too late. At one point comprising dozens of members, most of these self-proclaimed "immortals" had died of old age. The survivors seemed unfazed. "We still think some of us will make it," Elect Philip, a cadaverously thin old man, told the *Orange County Register* in 1995. At that point, the seven people left in the group—all octogenarians and nonagenarians—still believed the apocalypse was imminent.

I tracked them down by calling Charna Walker, the owner of Date-land, a nearby town. (Walker and her husband purchased the entire municipality—consisting of a gas station, a diner, a gift shop, an RV park, a water well and a date farm—in 1994.) "I'll be honest with you, I don't know a lot about them," said Walker, "and I don't want to say too much about them because I don't think it would be right." She did, however, provide a phone number.

Someone named Elect Star picked up. I told her about this book, and asked if I could come by and interview them. "Sure," she said, laughing, "I guess that would be fine."

Three weeks later, Liane and I drove to Dateland, passing through Arizona's scorching scrub desert. Hundreds of tumbleweeds rolled by like clusters of brittle bones, or the aimless remains of dried-up sea corals. After stopping for a date shake at the Walkers' oasis, we turned onto a forlorn country road.

There was no sign of humankind for miles. Eventually a derelict saloon appeared, seemingly sinking into the ground. Faded cursive letters spelled out its name: "The Whispering Sands." We drove on through the desolate landscape for another twenty minutes, when an abandoned, sagging church materialized. It looked like the ruins of an ancient civilization that had succumbed to a rapid, mysterious decline. I imagined the pews lined with repentant tumbleweeds, hungover from a hard night at the saloon.

We kept going. The only signposts were for gulches with names like "Hoodoo Wash." Just as I was beginning to wonder whether all of this was a mirage, an eight-foot-tall lollipop appeared on the horizon. In the middle of its rainbow swirl, hand-painted block letters announced CHILDREN OF LIGHT: 1 1/2 MILE AHEAD.

We pulled to a stop in the sandy driveway next to a mid-century bungalow with a large stone chimney and oversized windows. There was a flag above the house, emblazoned with a golden star reading "Purity, Promise, Peace, Perfection." As we shut our car doors, marveling at the many date palm trees, three seniors dressed identically in white robes, red vests and blue aprons came out to meet us. Their names were embroidered on their vests: Elect Star, Elect Philip and Elect David. "There's just three of us now," said Philip, explaining how there had once been more than sixty Children of Light.

Elect Star went into the sewing room to get us a crayon-colored diagram that explained how the Children of Light were supernaturally chosen by God in 1949.

Philip brought us to a painting he'd done of many different fruits on the same tree. Pointing at the trees, he explained that they grow a variety of Old World dates, such as medjools, halawis, khadrawis, barhis and dayris.

Philip led us into an oval prayer room, and began to relate the sect's history. They had started out as fruit farmers in Keremeos, British Columbia, a small town known as "The Fruit-Stand Capital of Canada." Located in a fertile valley surrounded by mountains, Keremeos attracted international attention in January 1951 after the Children of Light shut themselves onto a ranch, declaring that the world was about to end. "We put that town on the map—in disrepute," recalled Philip.

The group's leader, Grace Agnes Carlson—or Elect Gold, had been informed by God, in the form of a ball of fire rolling down the mountainside, that the world's expiry date was December 23, 1950.

That night, the congregation trekked in a procession, chanting hymns in the wind, to the top of "K mountain," so named because landslides had emblazoned it with an enormous letter K. Elect Gold said it stood for the Kingdom, which was at hand.

Judgment Day never came. When Christmas morning rolled around, the group descended the mountain and bolted their doors, awaiting further instruction.

Two earthquakes passed through the valley that week, rattling windows and dishes but causing no harm. Early in the New Year, word spread throughout the Okanagan Valley, and eventually to the news desks in Vancouver, that forty people, including schoolchildren, were holed up in a building waiting for the apocalypse.

Journalists staked out the ranch, making much of the fact that the group wore red capes lined with gold satin and white shirts intended to repel atomic rays. The papers also ran photos of a metallic bread loaf, believed to be one of the sect's devotional objects. If nothing else, the media concluded, congestion in the house constituted a danger to public health. As Philip walked us through the living room, he explained that even though they owned a television—which he called "hell-evision"—they used it only to watch Olympic skating videos or old Shirley Temple movies. "Everything's so dirty in these latter days," he sighed. "We try to watch things that are halfway clean."

In Keremeos, the media coverage led to unsuccessful raids by out-raged locals. Townsfolk went on tirades about hypnotized youths having unnatural experiences with grown-ups. A week later, the Mounties came in and removed some preteens from the compound. The kids emerged, scowling at the surrounding press corps.

On January 13, the group received a crucial sign: a cloud shaped like a human hand had changed from white to red to white again. That night, they tiptoed through the swirling snowfall, past dozing newsmen and photographers. Getting into their cars, they raced away, never to return. The blizzard caused an avalanche, which closed the pass off, so the media couldn't follow.

The Children of Light roamed around North America for the next twelve years, seeking their promised land. While in San Bernardino, they saw giant flaming letters in the sky spelling out "Agua Caliente, Arizona." Soon after that, a green disk appeared in front of the sun inscribed with the date "May 21, 1963."

On that very day, they arrived at their new home in Agua Caliente, this patch of Arizona desert not far from Dateland. It isn't quite clear how they came into possession of these eighty acres. Elect Star said, "We'd been told that God was going to put us on land of His ownership, and this was it: Agua Caliente." Philip told me a convoluted story involving hallu-cinations, a wealthy patron and a land-owning Native American named Scout Gray Eagle who'd dreamed of their coming.

However it happened, they've been here ever since. "Anyone is wel-come to come and stay for as long as they want," said Philip. As dinner was being prepared, we met two other elderly ladies who had moved in. They weren't full members and didn't wear the Children of Light uni-form. One of them was convinced that she'd met me in a previous life. She told me that hippies used to travel here in the late 1960s, but didn't stay long due to the sexless, drug-free atmosphere.

As for rock 'n' roll, the Children of Light write their own kinder-garten-like hymns, which they sang before dinner. At the kitchen table, everybody's name was written on folded cards indicating their place. Men sat opposite women, in keeping with the group's celibacy. Elect Star plugged in a child-sized Yamaha keyboard and the group pro-ceeded to serenade us with songs of praise. We joined in on a couple of choruses.

They explained their belief system—eat fruits, abstain from flesh relations and live forever—over a dinner of six dates, a dozen pistachios and walnuts, a smear of tree sap and some sliced canned peaches, all doled on sectional trays. The main dish was a peeled banana, which we ate with a fork and knife.

Although the group aspires to fruitarianism, they also eat homemade yogurt, popcorn and vegetables that grow on the farm. An angel, Philip said, once appeared to them and directed them to follow "the Garden of Eden diet," which is what we seemed to be eating that night. The angel also told them that it's a sin to be paid for work. "We haven't earned a penny in fifty-six years," said Philip. The group's members do, however, collect Social Security checks and they file every year with the IRS.

As I finished my last forkful of banana, I wondered what would happen when these virgins and eunuchs passed on. They were so old. Elect David, who ate with quivering hands, seemed to be on his last legs. They didn't seem too concerned about their organization dying out. After all, the world itself was dying out. A piece of paper pinned to a bulletin board read, "This is the most serious hour in the entire history of the world, for the Signs all declare that this is the time of the end on God's clock." On it, capitalized and underlined, was their motto: "ONLY THOSE WHO ATTEMPT THE ABSURD . . . ARE CAPABLE OF ACHIEVING THE IMPOSSIBLE!"

After dinner, we were invited to spend the night, or, if we wanted, the rest of our lives. Shaking hands with the three elects and their two friends, we thanked them for their hospitality and their stories, got into our car and waved good-bye.

By then, it was quite dark out. There were no streetlights, and the road was bumpier than we remembered. A minute or two later, when we reached the lollipop sign, Liane looked out her window and cried out: an enormous fire was raging near the compound. We considered turning around, but were too freaked out. Driving on, we called them on our cell phone to make sure they were safe. Elect Star picked up. I told her that a massive bonfire appeared to be burning right outside their building. She went out to look but came back to the line a moment later, claiming that she saw nothing. For miles, we kept catching glimpses of the fire glowing in the distance. Perhaps it was one of those mysterious wildfires that inexplicably burst to life in remote desert regions, or some sort of pyrotechnical ploy staged by the Children of Light to recruit us or maybe it was a sign from God. We didn't stick around to find out.

FRUITS HAVE PLAYED a role in other spiritual movements, from a Russian sect of berry-cultivating flagellants called the Krillovnas to John James Dufour's grape-growing colonies in Kentucky and Indiana. In the early twentieth century, Southern California's Societas Fraterna was led by a fruit Svengali named Thales. Also known as the Placentia Grass Eaters, the colony believed that spirits gathered in corners, so they built a mansion without any square rooms. Reports circulated of ghosts haunting the property, and balls of fire were sometimes seen shooting out of their chimney. The Societas Fraterna eventually disbanded after child-malnutrition lawsuits and the suicide of one of the group's young female members.

In the 1960s, nudist fruit-growing communes started popping up across America, as followers of the "back to the land" movement shed their clothes and inhibitions and banded together in utopian groups. In *Paradise Fever,* Ptolemy Tompkins writes of his own father's clothing-optional orchard community: "A key aspect of getting the garden going turned around one's being naked, or at least topless, while working inside its walls." (In *The Secret Life of Plants,* his father also once wrote that, "Apples experience the equivalent of an orgasm when eaten with a loving and respectful attitude.") Vestiges of those experiments can be found in biodynamic free-love farms and nudist gardens in hippie enclaves around the world. Mulberry grower and historian of naturalism Gordon Kennedy told me that in Nimbin, Australia, bush- and teepee-dwelling modern primitives called "ferals" still grow fruits in the nude.

As we've seen, many religions use fruits to represent the possibility of attaining divine consciousness, of entering into a communion with the infinite. Beyond the symbolism, fruits can literally affect our molecular structures in opaque manners. Many stimulants, like coffee, start out as fruits. Africa's kola nut is still used to make Coca-Cola and other colas. The areca nut, fruit of the betel palm, is a main ingredient in *paan,* the stimulant chewed in South Asia that stains streets crimson when spat out. Such energy-boosting fruits can become habit-forming.

The taste of certain delicious fruits can instill in the eater an instinctual urge to bow down before the majesty of the sensation. It took me years to decode the impulse, but I came to think of it as a *horizontal* feeling. It's so close to perfection that there's something deathly about it. At the same time, that sense of surrender, I think, is a momentary acceptance of our oneness with nature.

All over the world, fruits have been imbued with secret powers. Buddha, at the Great Miracle of Sravasti, made a mango tree grow instantly out of a seed. In India, rather than reading palms or tea leaves, fortune-tellers decrypt the future by observing flies landing on mango seeds. The Nupe tribe of northern Nigeria use strings of berries for divination. The Yoruba use palm fruits, and the Yukun use calabash disks.

In tribal societies, those who seemed capable of understanding the mystical nature of plants were anointed as shamans, messengers between this world and the other. For centuries, medicine men have used fruits to induce trancelike states. Any part of a plant can have entheogenic (god-releasing) aspects—roots, bark, resin, leaves, twigs, flowers or vines—but fruits seem particularly potent.

Ucuba fruits, called semen of the sun, cause tryptamine freakouts. Half a dozen Hawaiian wood rose seeds induce a state similar to an LSD trip. The potent datura fruit has been used in sacred rites since prehistoric times. The chilito is an acidic, edible fruit that grows on a Mexican cactus called *Hikuli mulatto*. It allows go-betweens to communicate with the beyond and is said to cause insanity in evil people, some of whom throw themselves from cliffs to escape the madness. The fresh fruits of the latúe facilitate dream work by Chile's Mapuche witch doctors. Mexicans crush ololiuqui fruits into a beverage drunk on quiet nights to create a potent hallucinogen. The red testicle-like fruits of the sanango, which grow in West Africa, cause visual hallucinations. The fruits and seeds of mandrake and henbane contain mind-bending alkaloids that were used in medieval witch brews. The fruit capsules of Sryian rue cause the same effects as ayahuasca. Cabalonga blanca fruits protect ayahuasca users from supernatural menace. Cebíl, the "seeds of civilization," are used to access another level of reality. Smoking banana peels, said to be hallucinogenic, is a rite of passage that usually ends in a headache. The Waiká of Venezuela use the beans of yopo fruits in a bizarre ritual where one man blows the fruit snuff through a long pipe into another man's nose. When he snorts it up, pandemonium ensues.

The psychotropic quality that many fruits possess is actually a form of toxicity. Plants contain poisons intended to deter small animals from messing with them. When humans ingest these plants, their neurochemicals have effects ranging from relaxation and inebriation to paralysis and death.

Cashew nuts are highly toxic until roasted. The seeds are surrounded by double shells that contain a skin irritant similar to poison ivy. This caustic liquid, called Cashew Nut Shell Liquid (CNSL), must be removed in order to eat the nut. Ackee, the national fruit of Jamaica, is tricky: when underripe, it contains hypoglycin, a violent purgative that can make you vomit until you die. Its seeds are always poisonous. Cyanide and cyanogens are found in citrus and pome fruit seeds. Starfruits contain oxalic acid. White sapotes are called "kill-health" (*matasano*) in Guatemala because of their narcotic seeds. Eating unripe monstera deliciosas is like biting into glass. Gambians dip their arrows into *Strophansus* fruits, which contain such virulent poison that it can kill a human in fifteen minutes. *Buah keluak* ("the fruit which nauseates"), used on spears, is eaten in Singapore after being buried underground for a month or more, and then soaked in water for weeks before being boiled. Orangutans eat and disperse the *Strychnos ignatii,* a fruit full of strychnine. Somehow immune to the poison, it just makes them salivate excessively.

"How I Almost Gave My Life in the Pursuit of Fruit" is the title of an article in *Fruit Gardener* magazine about an encounter with toxicity. The author, "Boobus Anonymous," starts by explaining her fantasy to become the discoverer of a delicious new fruit. She recounts how she noticed a sausage fruit tree growing on the campus of UCLA. Plucking one of the fruit, which look like large, dangling Hungarian salami, the author gave it a nibble, noting that it tasted like damp cornstarch. Realizing it wasn't the fruit that would earn her renown, she forgot about it, until, half an hour later, she noticed a tingling in her face.

Her mouth, she became convinced, had turned into a snout. Soon enough, she was fully hallucinating: "Reality had become an enormous transparent rubber band, stretched and encompassing the entire range of my vision, indeed my entire consciousness stretched out and distended, and I was inside it with no immediate options for getting out."

She put on nice clothes in case she died. Trying not to panic, she called some other fruit connoisseurs who helped her calm down, and informed her that she wouldn't die from the toxins. "Nowadays, when I walk past that tree and look at its hard, fat brown fruits," she concludes, "I feel a kinship—a link with that tree. Mentally I whisper, 'I know you . . . I've experienced you, probably as no one else has. And it's our secret. No one else suspects what you're really like.'"

WE DIE FOR THEM, we make love to them and we use them to contact the divine. We can be so entranced by fruits that we end up needing them to the point of self-destruction. Winos are addicted to fermented grapes, and heroin addicts are addicted to sap of the poppy fruit. In Samuel Beckett's *Krapp's Last Tape,* the narrator believes himself to be addicted to bananas, and in one of his recordings vehemently admonishes himself to "Cut 'em out!"

History's most excessive pomophile was Emperor Claudius Albinus, who ate ten melons a day, alongside five hundred figs, one hundred peaches and mounds of grapes. Emperor Diocletian abdicated his reign in order to dedicate himself to his beloved fruit trees. Antoine-Girard de Saint-Amant was a baroque poet who *really* loved melons: "Ha! Hold me up, I swoon! This delicious morsel tickles my very soul. It oozes a sweet juice which will steep my heart in ecstasy . . . O far better than gold, O Apollo's masterpiece! O flower of all fruits! O ravishing MELON!"

Prominent historical figures were believed to have perished due to their yearnings for edible fruit flesh, particularly for melons. Pope Paul II died in 1471, alone in his room in a melon-induced apoplexy. In 1534, a surfeit of melons took the life of Clement VII. Both Frederick III of Germany *and* his son Maximilian II died of melon overdoses. Others who fell victim to the assassinating melon include Guy de la Brosse, Louis XIII's doctor; the Baron de Rougemont in 1840; and Albert II of Germany. King John died in 1216 after overdoing it with peaches.

Artists, always intense, have taken fruit appreciation to maddening lengths. Alexandre Dumas ate an apple every morning at dawn beneath the Arc de Triomphe. He offered his life's work to the town of Cavaillon in exchange for an annuity of melons. The Dadaist George Grosz said that his friends and he used to eat "gooseberries until our bellies swelled up like zeppelins and we lay there like the wanderers in the land of Cockaigne after they had eaten their way through the great cake mountain." Andy Warhol has written about a side effect to gorging on cherries: "You have all the pits to tell you exactly how many you ate. Not more or less. Exactly. One-seed fruits really bother me for that reason. That's why I'd always rather eat raisins than prunes. Prune pits are even more imposing than cherry pits." Hitchcock ate gooseberries, seeds and all, every morning. Coleridge liked to bite fruits straight off trees, without using his hands. Agatha Christie wrote in the bathtub, surrounded by green apples. Friedrich Schiller kept rotten apples inside his desk and would breathe in their decaying perfume for inspiration. D. H. Lawrence climbed nude into mulberry trees to write. Henri Michaux claims that he spent twenty years learning how to project himself into fruits: "I put an apple on my table. Then I put myself into the apple. How peaceful!"

I too went off the deep end trying to get to the core of fruits. Wanting to understand these fruit votaries' passion, I spent months combing through any book that contained the word "fruit" in index searches. I fell into a vortex of ancient fruit books at the library of the New York botanical garden, at the Royal Botanical Garden's library in Niagara and at the Los Angeles Central Library.

There are more than 3,500 books exclusively about fruits, and over 8,000 books mainly about fruits. While sifting through pomological treatises, I'd think of Sylvia Plath's description of reading about a man and a nun who would meet at a fig tree to collect fruits, until their hands accidentally brushed one day, and the nun never returned: "I wanted to crawl in between the black lines of print the way you crawl through a fence, and go to sleep under that beautiful big green fig tree." I started daydreaming that Julia F. Morton, author of the legendary *Fruits of Warm Climates,* had an extramarital affair with the fruit hunter Wilson Popenoe, author of *Manual of Tropical and Subtropical Fruits.*

I took out dozens of books in an attempt to find the angel at the center of the rind, as Wallace Stevens once put it. I tried to figure out what variety of lemons were held in the hands of self-immolating Indian widows as they transformed themselves into embers on the banks of the Ganges. When Litvinenko, the former KGB spy, was poisoned in London, I wondered why nobody had thought of giving him cornellian cherries, which can leach radioactivity from our bloodstream. I tried to find out precisely which fruits were eaten by the fabled lotus eaters, torn between speculation over whether it was the Libyan coast's lotus jujube (*Zizyphus lotus*) or the carob, whose fruit is sweeter.

I pored through the esoteric works of Jakob Lorber, a nineteenth-century German mystic who spent the last twenty-four years of his life in, as Jorge Luis Borges describes it, "a series of protracted revelations." Starting in 1840, the voice of God commanded Lorber to put pen to paper and transcribe what he heard. From that moment on, until he died in 1864, he wrote all day almost every day, filling twenty-five volumes of over five hundred pages each (not including his minor works). Lorber wrote about many of the fruits found in outer space. Saturn, I was over-joyed to learn, produces pyramid fruits, fire fruits and rainbow-colored ship fruits used as boats. Ubra fruits, Lorber said, are nine-foot-high mercury pouches that grow on branchless trees whose square trunks of green glass shine like mirrors, allowing passersby to check out their reflections.

There are so many examples of fruits in mythology you can spend the rest of your life documenting them; in fact, that's kind of what happened to James George Frazer, author of *The Golden Bough*. He set out to explain an odd ritual involving forest-dwelling priests who attained priesthood by murdering their predecessor after plucking a certain sacred branch. This fateful branch was the eponymous Golden Bough. In his round-about evaluation of what that ritual murder might have signified, Frazer's work ballooned to twelve volumes. Even the abridged version is packed with examples of fruits and their roles in arcane, magico-religious cere-monies and primitive beliefs. Indeed, Frazer concludes that the Golden Bough itself was probably a sprig of berries still used ceremonially today: mistletoe.

The fruit world is endless. There are countless subcommunities full of passionate aficionados, whether it's plant pathologists pursuing a par-ticular fruit disease or germplasm collectors scouring the globe for the putative progenitor of the papaya. "People are amused that I've picked such a narrow topic, but fruit connoisseurship is not a narrow focus," Karp told me. "Narrow is investigating rootstock for apples. And even that's not really narrow—that's seminarrow." Agrilitholigists (people who collect fruit-crate label art) pay thousands of dollars for images dat-ing back to the 1800s. The central node in this network is Pat Jacobsen, a collector who has spent over thirty years writing hundreds of thousands of words about fruit-crate labels.

I thought back to Robert Palter, with his never-ending literary fruit episodes, and to Ken Love, with his thousands of pages on loquats and to Graftin' Crafton Clift's jungle rooms and the Fruit Detective's fruit fixa-tion. They all seemed lost in a pursuit of the unattainable—as was I.

This obsession with fruits is the desire to somehow know it all, to become omniscient. Perhaps after tasting the fruit of the knowledge of good and evil, we might then turn to the other tree's fruits and find ever-lasting life.

But, as Genesis insinuates, attaining knowledge won't necessarily set us free. On the contrary, it may enslave or even kill us. A duty must be paid on the beauty. The search for forever is never-ending. We might lose ourselves in the symbolism. Utopia is Greek for "nowhere," yet, as Oscar Wilde wrote, a map without it isn't even worth glancing at. We can always, never go there.

The oldest written story is a forest quest for immortality. But what Gilgamesh finds instead is humanity, an understanding that the same ineluctable fate awaits us all: a return to the soil. Everything is born from the earth and dies in it; it is both womb and tomb.

The fantasy of creating some perfect world and inhabiting it forever fuels much artistic creation. But there comes a time when every monkey must let go of its paradise nut. "Heaven," wrote Emily Dickinson, sagely, "is what I cannot reach! The Apple on the Tree."

16

Fruition: Or the Fever of Creation

Talking of Pleasure, this moment I was writing with one
hand, and with the other holding to my Mouth a
Nectarine—good God how fine. It went down soft
pulpy, slushy, oozy—all its delicious embonpoint melted
down my throat like a large beatified Strawberry. I shall
certainly breed.

—John Keats, 1819, private correspondence

IN THE MID-1800S, when delicious pears first appeared in America,
New Englanders were smitten. "You should have seen one of the
old-time pear parties," wrote E. P. Powell in 1905, recalling smoking
rooms full of Bostonians groaning in pleasure and rubbing their hands
with glee as they gorged themselves on new creations. Those who didn't
experience the frenzy, said Powell, can hardly understand the thrill
caused by the sudden appearance of soft, sweet, nectar-filled pears. The
phenomenon, known as "pear mania," was the fruit world's equivalent of
the British invasion in pop music.

The Lennon and McCartney of this craze were two Belgian growers
from Mons. Nicholas Hardenpont and Jean Baptiste van Mons bred
beurré pears, like the Bosc and the Flemish Beauty, that had the texture of
softened butter filled with dripping juice.

Before their pioneering efforts, pears were divided into two categories: those that tasted like shampoo, and those that didn't. Although Pliny documented forty-one varieties of pears in the first century A.D., he also explained that they were indigestible unless boiled, baked or dried. Darwin noted that pears were of very inferior quality in classical times. Dry, sandy and gritty, they were primarily used in America to make perry, a pear cider.

Until the Renaissance, juicy pears were almost inconceivable. They were so rare that only kings had access to varieties like the *Ah! Mon Dieu,* which is what Louis XIV purportedly ejaculated upon tasting it. It was only in eighteenth- and nineteenth-century Belgium, with the discoveries of Hardenpont and van Mons, that good pears hit the mainstream.

These new fruits found their most receptive audience in the New World, writes Ian Jackson in his unpublished *History of the Massachusetts Pear Mania of 1825–1875.* High-society tasting parties led besotted investors to throw capital into speculative orchards, many of which flopped. "There has been more money lost than made, for I could enumerate five persons who have utterly failed to every one who has made pear culture profitable," wrote P. T. Quinn in 1869's *Pear Culture for Profit.*

The Massachusetts Horticultural Society's autumn fruit shows were riotous affairs. There was fierce competition over novelties such as the Anjou, the Sheldon and the Claigeau. The possibility of tasting these pears, and getting rich growing them, galvanized legions of American amateurs to experiment with their own varieties. Everyone from orchard-owning factory bosses to workers with suburban plots of land suddenly got involved in pears. In a letter to Massachusetts growers, van Mons explained that the best was yet to come: "If you bravely persevere at raising seedlings you will end up with better than mine." His method was simple: plant out the seeds of juicy pears and hope for even juicier fruits to arise.

Unfortunately, most pear trees grown from seed bore unpalatable fruits. Ninety-nine out of a hundred trees can be tossed, cautioned A. J. Downing in his 1845 book *The Fruits and Fruit Trees of America.* Still, to anyone curious about fruit, he wrote, "nothing in the circle of culture can give more lively and unmixed pleasure than to produce and create—for it is a sort of creation—an entirely new sort."

Well before Darwin published his theories, fruit growers were using artificial selection to create superior fruits. In the wild, fruits need merely be healthy enough to ensure that the seeds reproduce. Fine eating fruits all stem from human cultivation. The pioneers behind the varieties we enjoy today are usually anonymous. As Jacques-Henri Bernardin de Saint-Pierre wrote, "The names of these public benefactors are chiefly unknown, whilst their benefits pass from generation to generation; whereas those of the destroyers of the human race are handed down to us on every page." The North American Fruit Explorers proudly describe themselves as the spiritual heirs to the fruit experimenters of yesteryear: "men and women, both known and unknown, who—in every age, the world over—have labored to discover, to adapt, and to improve the Best in Fruit."

Darwin, in *The Origin of Species,* acknowledged the role our ancestors played in ameliorating our fruits, noting how they produced splendid results from poor materials. "The art, I cannot doubt, has been simple. It has consisted in always cultivating the best known variety, sowing its seeds, and, when a slightly better variety has chanced to appear, selecting it, and so onwards."

The Polish poet Zbigniew Herbert once composed a prayer that read, in part: "Lord, Help us to invent a fruit / A pure image of sweetness." Lamentably, the twentieth century's emphasis on aesthetics and quantity led to many inferior tasting fruits. Huge, hard fruits with prolonged shelf lives were just fine for the shippers, wholesalers and retailers—but now breeding is being done with customers in mind. Flavor is just one of numerous variables that is taken into account when fruits are selected for breeding. Others include appearance, durability, shelf life, yield, size, shape, color, resistance to pests, flowering date and amount of bloom, crop volume and the ability to harvest before maturity. Most fruits aim for some degree of tastiness, but achieving it alongside the jolt-proof requirements of mass transportation has proven elusive—thus far.

BREEDERS AROUND THE WORLD, whether governmental agencies, university research labs or private companies, are all working on natural ways to develop better fruits. New Zealand's biochemistry firm Hort-Research will be making a fortune in coming years as their red-fleshed apple—a revamped heirloom—hits the mainstream. The company's star fruit breeder, Allan White, a self-styled "fruit fashion designer," has also created a non-GMO Bartlett pear and Asian pear hybrid that he says tastes so good it'll blow people's socks off.

Quebec government fruit breeder Shahrokh Khanizadeh recently found a type of apple that doesn't turn brown after being sliced open. The fruit wasn't produced through genetic engineering; it was an all-natural mutation that just happened to turn up in his orchard one day. The Eden maintains its whiteness, freshness and flavor for a week after being sliced. Everyone from McDonald's to major supermarket suppliers have been flying in to investigate his discovery. It could become an all-natural alternative to preservative-coated apple slices.

As we've seen, edible fruits are human artifacts. They will certainly continue to improve—if we want them to. To think otherwise is to misinterpret evolution. Some people think that heirloom fruits are invariably better; others believe that the hubbub over ancient varieties is sentimentality. "I'm not turned on by old apples like the Esopus Spitzenberg or the Ribston Pippin—they all have faults," says Phil Forsline, curator of the USDA's apple germplasm repository in Geneva, New York. He is surrounded by thousands of the best apples known to mankind, yet his favorite apple is the Honeycrisp, a new variety created by the University of Minnesota. It's large, juicy, has a satisfying crisp texture, lasts forever and tastes excellent.

The best raspberry I've ever tasted, the Tulameen, is an enormous gem that was bred in the 1980s by Hugh Daubeny at Canada's agricultural department. Mikeal Roose of the University of California at Riverside has created new seedless mandarins like the Yosemite, which tastes as sweet as Kool-Aid. And one of the finest strawberries available today was only developed in France in 1990. The Mara des Bois, as it is called, is the result of a quest to create a well-sized strawberry with the aroma and flavor of tiny wild strawberries. It is a blend of four cultivated varieties—Gento, Ostara, Red Gauntlet and Korona—none of which are wild, but all of which have prominent flavor characteristics.

One Southern Californian strawberry grower has ambitious plans to replace the rubbery strawberries we eat today. "The first time I tasted a Mara des Bois," says David Chelf, "I was really taken back by the perfume that explodes when you bite into it. It just penetrates your sinus cavity, your olfactory glands—wow. I loved strawberries as a child, but this exceeded my memory of the best strawberry I'd ever had as a kid."

Having tasted a box of his Mara des Bois, I can only say that I agree. They're a major leap from what most of us have access to currently. Chelf's company, Wicked Wilds, hopes to make flavorful strawberries widely available. "To let the cat out of the bag, I've set up a technique that can be reproduced around the world," he says. "Whether you're in New York or in London, it is now possible to grow organic strawberries year-round that actually have flavor."

His plan involves a patented greenhouse that uses simple aluminized mylar reflectors to intensify the light and runs about a quarter of the cost of a traditional greenhouse. This structure could be set up easily in the environs of any urban center, making local strawberries growable even in otherwise unhospitable environments. He is also fine-tuning a simple solar-powered version for use in developing nations. Additionally, these berries are all organic, because the greenhouse eliminates most pest concerns. "We had an aphid infestation recently, so I dumped five dollars worth of ladybugs in the greenhouse. They ate all the aphids within a couple of days. Even low-risk chemicals should be avoided. Let's use ladybugs!"

Another young maverick, Cornell researcher Jocelyn K. C. Rose, is developing new ways to increase the shelf life of fruits. "Pretty much all the fruits we eat are picked at the green stage," explains Rose. "They haven't had time to ripen properly, which is why they taste like biting into mothballs." He is attempting to find ways of growing fruits that have all the flavor components we associate with ripeness—minus the softness that keeps fruits out of the distribution cold chain. "If we are successful," he says, "it will be the holy grail of postharvest fruit biology."

Previous research has focused on the cell walls inside fruits. Unfortunately, those cell walls are critical to flavor and texture; strengthening cell walls usually leads to mealy, insipid fruits. Rose thinks there's another solution: strengthen the fruit's skin.

Rose has found a naturally occurring European tomato that stays ripe and firm and keeps its shape for six months. If he can understand how the proteins in the tomato's skin function, Rose believes he'll have found a way for the beneficial traits we associate with heirloom fruits—high acids, high sugars, great flavor, all the important nutrients—to withstand the rigors of commercial packing.

While Rose investigates tough skin, another breeder has been doing ambitious, long-term research that is already significantly enhancing the flavor of our fruits. Floyd Zaiger invented the pluot, that delicious—and immensely successful—plum-apricot hybrid. The most important fruit hybridizer of modern times, Zaiger has spent decades creating flavorful stone fruit intended to survive the cold-chain paradigm.

To find out where our fruits are headed, I visit the laboratories of "The Family Organized to Improve Fruit Worldwide."

WHEN I ARRIVE at Zaiger Genetics' research facility in Modesto, California, the front door is nudged open by a friendly leopard-print dog with a furry black back, a gray wolf neck, auburn thighs, a Dalmatian's flank and terrier ears. "He's a real Heinz 57," chuckles Floyd Zaiger, a smiling man in his eighties wearing a baseball cap and overalls. Inside the office, which is more like a forest ranger's cabin than a breeding laboratory, I meet Zaiger's grown-up sons, Gary and Grant, and his daughter, Leith. Floyd explains that his late wife had liked the name Keith and modified it for his daughter.

Floyd's fascination with unexpected juxtapositions goes back to his childhood, when his favorite food was tinned salmon with bananas. "You could buy a bunch of bananas and two salmon tins for a quarter," he says. "I loved eating them together."

Pluots contain more plum that apricot, but the Zaigers have also bred a series of interspecifics that are more apricot than plum, called apriums. He's also created peach-plums, spicy nectaplums and peacharines. The nectacotum is a nectarine, apricot, plum mix that sounds like a Norwegian black metal band. His peacotum, a blend of peach, apricot and plum tastes like fruit punch. "It can go kiwi," says one Zaiger salesman, suggesting it could possibly hit the big time.

Zaiger is also the reason we're able to buy white-fleshed peaches and nectarines. Until Zaiger figured out how to make them transport-ready, delicious white stone fruits were too soft to withstand any sort of shipping. Today, close to a third of peaches and nectarines sold are white-fleshed. When harvested at the right time, Zaiger-developed white-fleshed stone fruit are superlative.

The family's next big thing will be cherries crossed with plums. (Chums?) These aren't the small cherry-plum hybrids that occur in nature, like the feral myrobalan: Zaiger is talking about Bing cherries the size of a large plum.

Such mutations aren't genetically engineered using DNA splicing; they are bred naturally. Hybrid fruits are made by dusting pollen from the stamen of one flower onto the pistil of another and then growing out the resultant seed. This approach is limited: only flowers from related species can pollinate each other. A peach blossom can interact with another peach blossom to create a nectarine; but pineapple pollen, say, won't have any effect on the stigma of a kumquat.

Wild hybrids arise when the seed of a cross-pollinated fruit grows into a new tree. The loganberry, a hybrid of a blackberry and raspberry, just appeared one day in the backyard of a Judge J. H. Logan of Santa Cruz. *Prunus* fruits like plums, apricots, peaches and cherries all have an ancestral proto-parent that originated somewhere in Central Asia, meaning that whenever their flowers commingle, they yield interspecific hybrids. Melon flowers hybridize freely among themselves. One haphazard fusion of cantaloupes, honeydews and banana melons is called the "Cantabananadew."

What can't a banana do? Hybridize with a melon, unless transgenic engineering is used. Zaiger, who does no DNA splicing whatsoever, has devoted his life to crossbreeding *Prunus* flowers. He takes male pollen from as many different flowers as possible and mixes them with the female stigmas of as many other flowers as possible, and then plants the seeds to see how the hybrid expresses itself. Every year, he plants fifty thousand crosses, hoping to yield a handful of successes like pluots, speckled dinosaur eggs or apriums.

As Floyd wraps up a meeting, Gary walks me out of the office and into a large greenhouse full of trees in bloom. It smells like sweet perfume. Festive mariachi music is playing in the background. The floor is cushioned with petals. A dozen or so women are up in the trees, decorticating blossoms. Gary, whose hair is covered in flower parts, holds up a plum flower to show me its stamens. Yanking them off, he explains that the team works with Zaiger-designed tweezers to pluck and remove the male parts of every flower. This flower emasculation is done in order to expose the pistil. The castrated stamens are then snipped up with scissors. To collect the pollen, which looks like golden dust, the diced anthers are sifted through tea strainers. The pollen is then delicately dabbed onto the naked pistils of other fruit trees by women on ladders using Walgreen's eye-shadow brushes. "You have to be very patient to do this work," says a woman in a bright pink floral shirt. "Men can't do it."

Floyd comes over and gazes happily at the ladies fluttering about on ladders. "They're imitating the birds and the bees," he laughs deeply, I'm struck by his uncanny resemblance to Hugh Hefner. This isn't a mere fruit and flower factory; I'm in the Playboy mansion of the fruit world.

IN THE LATE eighteenth century, Goethe noticed how fruits arise as flowers "die into being." At that time, fruits were often grown by prisoners and monks. As 1768's *The Fruit Gardener* put it: "men cut off from society must have amusements." Like other monastic shut-ins, Gregor Johann Mendel amused himself by growing plants. His particular fancy was peas. As he dabbled with them, he made a momentous realization. By mixing the pollen from one flower with the stigma of another, he was able to create all sorts of hybrids. After growing close to thirty thousand pea plants in his experimental garden in Moravia, he presented a paper called "Experiments on Plant Hybridization." It was thoroughly ignored. Mendel died unknown and unrecognized.

In the early twentieth century, however, his work was rediscovered. Mendel was posthumously baptized the "father of modern genetics." His work shed light on the mysteries of heredity, paving the way to the mapping of the double helix and the pioneering research molecular geneticists are doing today with DNA—some of which has led to public outcry.

Nine years after Mendel's death, in 1893, a California fruit grower named Luther Burbank published a catalog called *New Creations in Fruits and Flowers*. It contained fruit marvels such as a strawberry-raspberry hybrid; a cross between the California dewberry and the Siberian raspberry; and a freakish apple that tasted sour on one side and sweet on the other. His hybrid of the African stubble berry and rabbit weed yielded an entirely new form; although neither of the parent plants bore edible fruits, his hybrid bore a delicious berry. Burbank was dubbed "the wizard of horticulture" because of his ability to create new fruits. Plants and fruits, he said, are like "clay in the hands of the potter or color on the artist's canvas and can readily be molded into more beautiful forms and colors than any painter or sculptor can ever hope to bring forth." The condition of nature, he seemed to be saying, is absolutely unfinished. Fruits are continually evolving.

Burbank anticipated Zaiger's efforts by breeding the *Prunus salicina,* a plum intended to be shipped cross-country. Zaiger got his start working with a former Burbank apprentice named Fred Anderson. Known as "the Father of the Nectarine," Anderson bred the first commercial nectarines in America. "Fred got the breeding bug from Burbank, and I caught the dreaded disease from Fred," says Zaiger. While working under Anderson's tutelage, Zaiger started breeding cherry-apricot hybrids in search of trees that would be resistant to soil, climate and pests. At the outset, the trees were all sterile, but then some of them started bearing fruits. Tasting them was an epiphany that changed Zaiger's life—and our fruits.

Zaiger grew up during the Dust Bowl depression. "When I was a kid," he recalls, "the apples were mostly wormy and you were lucky at Christmas if you got an orange in your sock. We didn't eat any peaches because we couldn't afford them." Zaiger's lack of fresh fruits as a child triggered his life's work of making good fruits as widely available as possible.

"Today, we're breeding for the whole world," says Zaiger, not immodestly. He has sold millions of trees, which have produced untold billions of fruits in countries like Argentina, Chile, Brazil, Egypt, China, Tunisia and South Africa. The Zaiger family's goal is to create new fruits that work within the chains of distribution. "People in high-rises want high-quality fruits, but those fruits need to be shippable in order to reach them," explains Gary. "Peaches all used to be so fragile that a fingerprint would bruise it. We create varieties that have similar flavor-profiles as the old varieties, but that are bigger, firmer and can ship well. They're not just a bag of water anymore. A peach that'll melt all over you and drip onto your shirt is great, but they can't be sold. Some couldn't be carried across the road without denting. If you want to grow them yourself, that's great. We make varieties for the home gardener as well. And the best old varieties are still available as well."

The Zaigers are optimistic about the future. "Right now we have things that look good and ship well, but the flavor isn't quite there yet," acknowledges Floyd. "Within twenty years, it'll fully be there. Flavor is the number-one criteria we push for now. The whole gist of our program is to improve the industry. In the future, there will be stone fruits that taste delicious after three weeks of shipping. I'm sure they exist. If we work hard enough, it will be possible to find them."

Floyd shows me a diagram for one of the hybrids they've been working on. It's an agglomeration of dozens of different fruits. The separate varieties combine together in a rhizomatic scheme until the result, a Dinosaur Egg or Dapple Dandy, comes out at the end. He explains how it's important to break up linkages between genes, and how some genes may actually exist in fruits but may not be expressing themselves and how different DNA must be lined together in order to make hidden traits come to life.

"I think I understand," I say.

"I wish I did!" he answers. "The further you go the more you realize how little you really know. The wealth of information is ever-expanding. Learning about fruits is like entering a funnel from the bottom—as you get into it, it keeps getting wider and wider."

He then opens a drawer on his desk and pulls out an envelope. He hands it to me, saying, "I look at this whenever I get discouraged."

Inside is a letter, dated August 18, 2005, from a fan in Boise, Idaho: "I have recently become addicted to Pluots . . . My sincerest thanks to all of you for making these fruits available to me and my family. With such yummy fruit it's easy to get our five servings of fruits and vegetables per day—some days it's five pluots. We hope the vegetables forgive us."

LEAVING THE ZAIGERS', I point the rental car back toward San Francisco. The thought of Zaiger's funnel is strangely comforting.

It reminds me of something Stephen Wood, who grows heirloom apples at Poverty Lane Orchards, told me as we were munching Esopus Spitzengergs that autumn day. Despite working with apples for decades, he said that all he'd learned was how to grow a few apple varieties in his climate. "The more you know about fruits," he explained, "the more you realize how little you know."

It was Kierkegaard who noted how "there comes a critical moment where everything is reversed, after which the point becomes to understand more and more that there is something which cannot be understood." Fruits bring us to the brink of the eternal unknowable. Beyond it, the natural collapses into the supernatural.

I think I've learned enough. Enough to know that I can never know enough. Enough to remain awed by their infinitude. Enough to accept that I might never make it to Australia's outback to sample bush fruits like the red quandong, which is considered a gourmet treat, or the blue quandong, which looks like an ostrich egg painted metallic blue with a glitter gun, or even the silver quandong, which is so far ahead of its time that fruities will only be eating them in the distant future. Although part of me still wants to visit the North American Fruit Explorers Annual Fruit Showcase in a few months, if only to hear their tales of "Exploring the Caucasus Mountains for Wild Fruits," I already know what it will be like: endlessly fascinating.

I pull out a stick of peppermint chewing gum, and at that precise moment notice that I'm crossing a street called Blue Gum Drive. As I turn onto the freeway, my field of vision fills with orchards. There is no end. Rows upon rows of trees stretch to the horizon, like upturned souls frozen in the afterlife. Their angular, leafless branches palpitate with blossoms. Renewal! The flowers are as bright and white as shards of ice in the winter sun. Soon the fruits will be ripening.

Acknowledgments

THIS BOOK could not have come to fruition without Kurt Ossenfort, David Karp, Mireille Silcoff, Taras Grescoe, Jocelyn Zuckerman, William Sertl, Charles Levin, Cat Macpherson, Sarah Amelar, Anna deVries, Martha Leonard, Kathleen Rizzo, Amber Husbands and Miska Gollner. Thank you to all the fruit growers, preservers, lovers, scholars and sellers for taking the time to meet and speak with me. The Canada Council for the Arts provided early support. Michelle Tessler, my fantastic literary agent, also helped shape the narrative. Sarah Rainone and Amy Black at Doubleday provided invaluable structural insights. Nan Graham, Susan Moldow and Sarah McGrath at Scribner believed in this book and made it all possible. I am incredibly fortunate to have Alexis Gargagliano as my editor. My deepest gratitude to Liane Balaban and my family.

Further Reading

Ackerman, Diane. *A Natural History of the Senses*. New York: Random House, 1990.

———. *The Rarest of the Rare: Vanishing Animals, Timeless Words*. New York: Random House, 1995.

Anderson, Edgar. *Plants, Man and Life*. Boston: Little Brown & Co., 1952.

Ardrey, Robert. *African Genesis: A Personal Investigation into the Animal Origins and Nature of Man*. New York: Atheneum, 1961.

Armstrong, Karen. *In the Beginning: A New Interpretation of Genesis*. New York: Alfred A. Knopf, 1996.

———. *A Short History of Myth*. Edinburgh: Canongate, 2005.

Asbury, Herbert. *The French Quarter*. New York: Garden City, 1938.

Atwood, Margaret. *Negotiating with the Dead: A Writer on Writing*. Cambridge, U.K.: Cambridge University Press, 2002.

Barlow, Connie. *The Ghosts of Evolution: Nonsensical Fruits, Missing Partners, and Other Ecological Anachronisms*. New York: Basic Books, 2000.

Barrie, James Matthew. *Peter Pan: The Complete and Unabridged Text*. New York: Viking Press, 1991.

Beauman, Fran. *The Pineapple: King of Fruits*. London: Chatto & Windus, 2005.

Behr, Edward. *The Artful Eater: A Gourmet Investigates the Ingredients of Great Food*. Boston: Atlantic Monthly Press, 1992.

Borges, Jorge Luis. *The Book of Imaginary Beings*. London: Vintage, 1957.

Brautigan, Richard. *In Watermelon Sugar*. New York: Dell, 1968.

Bridges, Andrew. "Ex-FDA Chief Pleads Guilty in Stock Case." *Washington Post*. October 17, 2006.

Brillat-Savarin, J. A., trans. and annotated by M. F. K. Fisher. *The Physiology of Taste*. New York: Alfred A. Knopf (1825), 1971.

Broudy, Oliver. "Smuggler's Blues." www.salon.com. Posted January 14, 2006.

Browning, Frank. *Apples: The Story of the Fruit of Temptation*. New York: North Point Press, 1998.

Bunyard, E. A. *The Anatomy of Dessert: With a Few Notes on Wine*. London: Dulau, 1929.

Burdick, Alan. *Out of Eden: An Odyssey of Ecological Invasion*. New York: Farrar, Strauss & Giroux, 2005.

Burke, O. M. *Among the Dervishes*. New York: Dutton, 1975.

Burroughs, William S., and Ginsberg, Allen. *The Yage Letters*. San Francisco: City Lights, 1963.

Campbell, Joseph. *The Hero with a Thousand Faces*. Princeton, N.J.: Princeton University Press, 1968.

Chatwin, Bruce. *Utz*. London: Jonathan Cape, 1988.

Cooper, William C. *In Search of the Golden Apple: Adventure in Citrus Science and Travel*. New York: Vantage, 1981.

Coxe, William. *A View of the Cultivation of Fruit Trees, and the Management of Orchards and Cider*. Philadelphia: M. Carey, 1817.

Cronquist, Arthur. *The Evolution and Classification of Flowering Plants*. Boston: Houghton Mifflin, 1968.

Cunningham, Isabel Shipley. *Frank N. Meyer: Plant Hunter in Asia*. Ames, Iowa: Iowa State University Press, 1984.

Dalby, Andrew. *Dangerous Tastes: The Story of Spices*. Berkeley: University of California Press, 2000.

Darwin, Charles. *The Origin of Species*. London: Murray, 1859.

Daston, Lorraine, and Park, Katharine. *Wonders and the Order of Nature, 1150–1750*. New York: Zone Books / MIT Press, 1998.

Davidson, Alan. *Fruit: A Connoisseur's Guide and Cookbook*. London: Mitchell Beazley, 1991.

———. *A Kipper with My Tea*. London: Macmillan, 1988.

———. *The Oxford Companion to Food*. Oxford: Oxford University Press, 1999.

Davis, Wade. *The Clouded Leopard: Travels to Landscapes of Spirit and Desire*. Vancouver: Douglas & McIntyre, 1998.

De Bonnefons, Nicholas, trans. by Philocepos (John Evelyn). *The French Gardiner: Instructing on How to Cultivate All Sorts of Fruit Trees and Herbs for the Garden*. London: John Crooke, 1658.

De Candolle, Alphonse Pyrame. *Origin of Cultivated Plants*. London: Kegan Paul, Trench, Trübner & Company, 1884.

De Landa, Friar Diego. *Yucatan Before and After the Conquest*. New York: Dover, 1937.

Diamond, Jared. *Guns, Germs, and Steel: The Fates of Human Societies*. New York: Norton, 1997.

Didion, Joan. "Holy Water," in *The White Album*. New York: Simon & Schuster, 1979.

Downing, A. J. *The Fruits and Fruit Trees of America*. New York: Wiley and Putnam, 1847.

Duncan, David Ewing. "The Pollution Within." *National Geographic*. October 2006.

Durette, Rejean. *Fruit: The Ultimate Diet*. Camp Verde, Ariz.: Fruitarian Vibes, 2004.

Eberhardt, Isabelle, trans. Paul Bowles. *The Oblivion Seekers*. San Francisco: City Lights, 1972.

Echikson, William. *Noble Rot: A Bordeaux Wine Revolution*. New York: Norton, 2004.

Edmunds, Alan. *Espalier Fruit Trees: Their History and Culture*. 2nd ed. Rockton, Canada: Pomona Books, 1986

Eggleston, William. *The Democratic Forest*. New York: Doubleday, 1989.

Eiseley, Loren. *The Immense Journey*. New York: Random House, 1957.

Eliade, Mircea. *Patterns in Comparative Religion: A Study of the Element of the Sacred in the History of Religious Phenomena*. Translated by R. Sheed. London: Sheed and Ward, 1958.

————. *The Sacred and the Profane: The Nature of Religion*. Translated by W. Trask. London: Harcourt Brace Jovanovich, 1959.

————. *Myths, Dreams and Mysteries: The Encounter Between Contemporary Faiths and Archaic Realities*. Translated by P. Mairet. London: Harvill Press, 1960.

————. *Images and Symbols: Studies in Religious Symbolism*. Translated by P. Mairet. London: Harvill Press, 1961.

————. *Myth and Reality*. Translated by W. Trask. New York: Harper and Row, 1963.

————. *Shamanism: Archaic Techniques of Ecstasy*. Translated by W. Trask. London: Routledge and Kegan Paul, 1964.

Epstein, Samuel S. *The Politics of Cancer Revisited*. New York: East Ridge Press, 1998.

Evans, L. T. *Feeding the Ten Billion: Plants and Population Growth*. Cambridge, U.K.: Cambridge University Press, 1998.

Facciola, Stephen. *Cornucopia II: A Source Book of Edible Plants*. Vista, Calif.: Kampong, 1990.

Fairchild, David G. *Exploring for Plants*. New York: Macmillan, 1930.

————. *The World Was My Garden*. New York: Scribner, 1938.

————. *Garden Islands of the Great East*. New York: Scribner, 1943.

Fisher, M. F. K. *Serve It Forth*. New York: Harper, 1937.

Fishman, Ram. *The Handbook for Fruit Explorers*. Chapin, Ill.: North American Fruit Explorers, Inc., 1986.

Forsyth, Adrian, and Miyata, Ken. *Tropical Nature: Life and Death in the Rain Forests of Central and South America*. New York: Scribner, 1984.

Frazer, J. G. *The Golden Bough* (12 volumes). London: Macmillan, 1913–1923.

Freedman, Paul, ed. *Food: The History of Taste*. Berkeley: University of California Press, 2007.

Fromm, Erich. *The Heart of Man: Its Genius for Good and Evil*. New York: Harper, 1964.

The Fruit Gardener, publication of the California Rare Fruit Growers. 1969–present.

Frye, Northrop. *Creation and Recreation*. Toronto: University of Toronto Press, 1980.

Gide, André. *Fruits of the Earth*. London: Secker & Warburg, 1962.

Graves, Robert, and Patai, Raphael. *Hebrew Myths: The Book of Genesis*. New York: Doubleday, 1964.

Grescoe, Taras. *The Devil's Picnic: Around the World in Pursuit of Forbidden Fruit*. New York: Bloomsbury, 2005.

Guterson, David. "The Kingdom of Apples: Picking the Fruit of Immortality in Washington's Laden Orchards." *Harper's.* October 1999.

Healey, B. J. *The Plant Hunters.* New York: Scribner, 1975.

Hedrick, U. P. The collected works.

Heintzman, Andrew, and Solomon, Evan, eds. *Feeding the Future: From Fat to Famine, How to Solve the World's Food Crises.* Toronto: Anansi, 2004.

Heiser, Charles. *Seed to Civilization: The Story of Man's Food.* San Francisco: W. H. Freeman, 1973.

———. *Of Plants and People.* Norman, Okla.: University of Oklahoma Press, 1992.

Hennig, Jean-Luc. *Dictionnaire Litteraire et Erotique des Fruits et Legumes.* Paris: Albin Michel, 1998.

Hopkins, Jerry. *Extreme Cuisine.* Singapore: Periplus, 2004.

Hubbell, Sue. *Shrinking the Cat: Genetic Engineering Before We Knew About Genes.* Boston: Houghton Mifflin, 2001.

Huysmans, J. K. *Against Nature.* New York: Penguin, 1986.

Jackson, Ian. The uncollected works.

James, William. *The Varieties of Religious Experience: A Study in Human Nature.* New York: Modern Library, 1902.

Janson, H. Frederic. *Pomona's Harvest.* Portland, Ore.: Timber Press, 1996.

Karp, David. The collected works.

Kennedy, Gordon, ed. *Children of the Sun.* Ojai, Calif.: Nivaria Press, 1998.

Kennedy, Robert F., Jr. "Texas Chainsaw Management." *Vanity Fair.* May 2007.

Koeppel, Dan. "Can This Fruit Be Saved?" *Popular Science.* June 2005

Levenstein, Harvey. *A Revolution at the Table: The Transformation of the American Diet.* Oxford: Oxford University Press, 1988.

Lévi-Strauss, Claude. *Tristes Tropiques.* New York: Athenium, 1971.

McIntosh, Elaine N. *American Food Habits in Historical Perspective.* Westport, Conn.: Praeger Press, 1995.

Nabhan, Gary Paul. *Gathering the Desert.* Tucson, Ariz.: University of Arizona Press, 1985.

Manning, Richard. *Food's Frontier: The Next Green Revolution.* New York: North Point Press, 2000.

———. *Against the Grain: How Agriculture Has Hijacked Civilization.* New York: North Point Press, 2004

———. "The Oil We Eat." *Harper's.* February 2004.

Mason, Laura. *Sugar Plums and Sherbet: The Prehistory of Sweets.* Totnes, Devon: Prospect Books, 1998.

Matt, Daniel C. *The Zohar: Pritzker Edition.* Palo Alto, Calif.: Stanford University Press, 2004.

McGee, Harold. *On Food and Cooking: The Science and Lore of the Kitchen.* New York: Scribner, 1984.

McKenna, Terence. *Food of the Gods, The Search for the Original Tree of Knowledge.* New York: Bantam, 1992.

McPhee, John. *Oranges.* New York: Farrar, Strauss & Giroux, 1966.

————. *Encounters with the Archdruid*. New York: Farrar, Strauss & Giroux, 1971.

Mintz, Sidney W. *Sweetness and Power: The Place of Sugar in Modern History*. New York: Viking, 1986.

Mitchell, Joseph. *Joe Gould's Secret*. New York: Viking, 1965.

————. *Up in the Old Hotel and Other Stories*. New York: Pantheon, 1992.

Morton, Julia F. *Fruits of Warm Climates*. Miami: Florida Flair Books, 1987.

Musgrave, Toby et al. *The Plant Hunters: Two Hundred Years of Adventure and Discovery Around the World*. London: Ward Lock, 1998.

Nabokov, Vladimir. *Speak, Memory*. New York: GP Putnam and Sons, 1966.

Nestle, Marion. *What to Eat*. New York: North Point Press, 2006.

O'Hanlon, Redmond. *Into the Heart of Borneo*. New York: Random House, 1984.

Pagels, Elaine. *Adam, Eve and the Serpent: Sex and Politics in Early Christianity*. New York: Vintage Books, 1989.

Palter, Robert. *The Duchess of Malfi's Apricots, and Other Literary Fruits*. Columbia, S.C.: University of South Carolina Press, 2002.

Partridge, Burgo. *A History of Orgies*. New York: Bonanza, 1960.

Piper, Jacqueline. *Fruits of South-East Asia: Facts and Folklore*. Singapore: Oxford University Press, 1989.

Pollan, Michael. *The Botany of Desire: A Plant's-Eye View of the World*. New York: Random House, 2001.

Popenoe, Wilson. *Manual of Tropical and Subtropical Fruits*. New York: Hafner Press, 1974.

Quinn, P. T. *Pear Culture for Profit*. New York: Orange Judd, 1869.

Raeburn, Paul. *The Last Harvest: The Genetic Gamble That Threatens to Destroy American Agriculture*. New York: Simon & Schuster, 1995.

Reaman, G. Elmore. *A History of Agriculture in Ontario*. Toronto: Saunders, 1970.

Reich, Lee. *Uncommon Fruits for Every Garden*. Portland, Ore.: Timber Press, 2004.

Roberts, Jonathan. *The Origins of Fruits and Vegetables*. New York: Universe, 2001.

Roe, Edward Payson. *Success with Small Fruits*. New York: Dodd, Mead, 1881.

Roheim, Geza. *The Eternal Ones of the Dream: A Psychoanalytic Interpretation of Australian Myth and Ritual*. New York: International Universities Press, 1945.

Root, Waverly, and de Rochemont, Richard. *Food: An Authoritative and Visual History and Dictionary of the Foods of the World*. New York: Simon & Schuster, 1981.

Rossetti, Christina. *Goblin Market*. London: Macmillan, 1875.

Sarna, Nahum M. *Understanding Genesis: The World of the Bible in the Light of History*. New York: Schocken Books, 1972.

Schafer, Edward H. *The Golden Peaches of Samarkand: A Study of T'ang Exotics*. Berkeley, Calif.: University of California Press, 1963.

Schivelbusch, Wolfgang. *Tastes of Paradise: A Social History of Spices, Stimulants, and Intoxicants*. New York: Pantheon, 1992.

Schlosser, Eric. "In The Strawberry Fields." *The Atlantic Monthly,* November 1995.

————. *Fast Food Nation*. Boston: Houghton Mifflin, 2001.

Schneider, Elizabeth. *Uncommon Fruits & Vegetables: A Commonsense Guide*. New York: Harper & Row, 1986.

Seabrook, John. "The Fruit Detective." *The New Yorker,* August 19, 2002.

———. "Renaissance Pears." *The New Yorker,* September 5, 2005.

———. "Sowing for Apocalypse." *The New Yorker,* August 27, 2007.

Shephard, Sue. *Pickled, Potted, and Canned: How the Art and Science of Food Preserving Changed the World*. New York: Simon & Schuster, 2001.

Silva, Silvestre, with Tassara, Helena. *Fruit Brazil Fruit*. São Paulo, Brazil: Empresa das Artes, 2001.

Soulard, Jean. *400 Years of Gastronomic History in Quebec City*. Verdun, Canada: Communiplex, 2007.

Steingarten, Jeffrey. "Ripeness Is All," in *The Man Who Ate Everything*. New York: Alfred A. Knopf, 2002.

Thoreau, Henry David. *Wild Fruits*. New York: Norton, 2000.

Tinggal, Serudin bin Datu Setiawan Haji. *Brunei Darussalam Fruits in Colour*. Brunei: Universiti Brunei Darussalam, 1992.

Tompkins, Peter, and Bird, Christopher. *The Secret Life of Plants*. New York: Harper & Row, 1975.

Tompkins, Ptolemy. *Paradise Fever*. New York: Avon Books, 1998.

Tripp, Nathaniel. "The Miracle Berry." *Horticulture,* January 1985.

Visser, Margaret. *Rituals of Dinner*. New York: Penguin, 1993.

———. *Much Depends on Dinner*. New York: Collier Books, 1986.

Warner, Melanie. "The Lowdown on Sweet?" *New York Times.* February 12, 2006.

Weisman, Alan. *The World Without Us*. New York: St. Martin's Press, 2007.

Welch, Galbraith. *The Unveiling of Timbuctoo*. London: Victor Gollancz, 1938.

Whiteaker, Stafford. *The Compleat Strawberry*. London: Century, 1985.

Whitman, William F. *Five Decades with Tropical Fruit*. Miami, Fla.: Fairchild Tropical Garden, 2001.

Whitney, Anna. "'Fruitarian' Parents of Dead Baby Escape Jail." *The Independent.* September 15, 2001.

Whittle, Tyler. *The Plant Hunters: 3,450 Years of Searching for Green Treasure*. London: William Heinemann, 1970.

Whynott, Douglas. *Following the Bloom: Across America with the Migratory Beekeeps*. Harrisburg, Pa.: Stackpole, 1991.

Wilde, Oscar. "The Decay of Lying: A Dialogue." *The Nineteenth Century,* January 1889.

Wilson, Edward O. *Biophilia: The Human Bond with Other Species*. Cambridge, Mass.: Harvard University Press, 1984.

———, ed. *The Downright Epicure: Essays on Edward Bunyard*. Totnes, Devon: Prospect Books, 2007.

Index

in burrs, 26, 57
 dehiscent, 26–27
 human-aided, 29–30, 37
Seed Savers Exchange, 220
serendipity berry, 176, 177
sesame fruit, 22
 seed dispersal by, 27
Seychelles Islands, 109–22
 ancient history of, 114
 buried treasures in, 119–21
 fauna of, 112
 as Garden of Eden, 110–11
 tourism industry of, 115, 116
 Vallée de Mai forest reserve of, 111–15, 117
 see also coco-de-mer
smuggling, 8, 34, 67–68, 123–39
 by conservationists, 132
 convictions for, 125, 129, 130, 132
 drug, 130–31
 for exclusive germplasm, 125, 132
 in foreign countries, 129–30
 governmental crackdown on, 130–31
 by immigrants, 132
 of immigrants, 131–32
 by Jefferson, 84
 methods of, 132–35
 profits of, 126, 128–29, 132
 as video game theme, 129
 by wealthy collectors, 134–37
Snyder, Gary, 148–50, 152, 153–60, 216
spices, 23–24
sports, grafting, 69
star anise, 25
Stewart, Martha, 146
strawberries, 12, 15, 29, 98, 146, 191, 208, 210, 211, 236, 237–40
 achenes of, 23, 83
 genetically modified frost-resistant, 180–81
 musk, 229
 new varieties of, 257–58
 varieties of, 83, 84, 177, 207, 221, 238, 239, 240, 257
 white, 71–72
supermarkets, 7, 52, 53, 55, 74, 86, 154, 179, 186, 195, 205, 207, 208, 233
Supreme Court, U.S., 22
Swank, Hilary, 129
sweeteners, 61, 169–79
 artificial, 174–75
 see also miracle fruit; miraculin

tea plants, 23
Thai Banana Club, 60–61, 182
Thailand, 61, 81, 91–97, 108–9, 126
theft, fruit, 137–38
Thoreau, Henry David, 13, 137, 157, 209

tomatoes, 7, 22, 72, 181, 205, 258
transmigration of souls, 103–5
Trewia nudiflora fruit, 28
Tribulus terrestris, 56
tropical fruits:
 atavistic response to, 88
 best international sources of, 72
 commercial production of, 60, 61, 63, 74
 explorers greeted with, 82
 growers of, 60, 61–62
 as intolerant of shipping, 133–34
 legal importation of, 74
 named for fruit hunters, 83–84
 propagation of, 85
 racketeering and, 199
 sales of, 186

ucuba fruit, 248
United Fruit Company, 185

vanilla, 22, 30
van Mons, Jean Baptiste, 254–55
Verlaque, Richelieu, 121–22
vertical integration, 152, 204
Victoria, Queen of England, 133–34
Volcere, Exciane, 112–13, 114–15
Voon Boon Hoe, 80–81, 87, 89–90, 132

Wallace, Alfred Russel, 79
Washington, George, 51, 70
watermelons, 13, 25, 55, 133, 211
 available cultivars of, 54
 heirloom, 220
 seedless, 30
 square, 177
Waters, Alice, 146
Waugh, Evelyn, 53, 143
wheat, 22, 27, 29, 47, 185
Whitman, Angela, 62, 63, 73
Whitman, William F., 40, 60, 61–63, 66, 79, 80, 87, 177
Wild Fruits (Thoreau), 157, 209
William F. Whitman Tropical Fruit Pavilion, 60–61, 66–71, 74
Wilson, Edward O., 5, 15, 225
Wilson, Richard, 67, 130, 177–78
Wolfram, Klaus, 97
Wood, Stephen, 216, 218, 263
World Trade Organization (WTO), 126, 127–28

XanGo mangosteen juice, 165

yohimbe tree, 168

Zaiger, Floyd, 259–63
Zebroff, George, 217–20

About the Author

Adam Leith Gollner has traveled around the globe to report on the fruit underworld. He has written for *The New York Times*, *Gourmet, Bon Appétit* and *Good* magazine. The former editor of *Vice* magazine, he is also a musician.

This is his first book. He lives in Montreal.